THE BEGINNER'S GUIDE TO
food
AND
WINE

THE BEGINNER'S GUIDE TO *food* — *AND* — WINE

HOW TO USE THE RECIPES

Approximate preparation and cooking times are given for each recipe, with the following symbols where appropriate:

Ease (one fork) easy to prepare and cook (two forks) requires care (three forks) complicated

Cost (£) inexpensive (££) moderately priced (£££) for special occasions

Freezing (snowflake) freezes particularly well

Plan ahead (clock) can be cooked ahead of time; this symbol in the steps indicates that the dish can be prepared ahead to this point.

Weights and measures Both metric and imperial measurements are given, but are not exact equivalents; work from one set only. Use graded measuring spoons levelled across.

Ingredients • Flour is plain white flour unless otherwise stated • Sugar is granulated unless otherwise stated • Eggs are medium (EEC size 4) unless otherwise stated • Vegetables and fruits are medium-sized and prepared, and onions, garlic and root vegetables are peeled unless otherwise stated • Butter for sweet dishes should be unsalted • Block margarine may be substituted for butter, although the results may be different in flavour, appearance and keeping qualities.

Cooking Dishes should be placed in the centre of the oven unless otherwise stated.

The following symbols represent the price of the selected wines (at the time of going to press): (one glass) under £3 (two glasses) £3-£5 (three glasses) over £5

This edition published in 1994
by Marshall Cavendish Books
(a division of Marshall Cavendish Partworks Ltd)
119 Wardour Street
London W1V 3TD

ISBN 1 85435 717 4

Some of this material has previously appeared in the
Marshall Cavendish partwork *WHAT'S COOKING?*

Printed and bound in the Slovak Republic

contents

foreword

AS VINE CULTIVATION has become more advanced and widespread, so the drinking of wine has increased in popularity. However, for the novice the world of wine can be daunting. What sets the large range of varieties apart? Which varieties grow where, and how successfully? Which are the new wines to look out for? Which was the best year for certain classic wines? And, as wine should be enjoyed with good food, which dishes bring out the best in which wines?

The Beginner's Guide to Food and Wine provides a clear, easy-to-follow introduction to 68 readily available popular and classic wines, with advice on what to buy, and when. In addition, each wine is accompanied by a recipe designed to complement its unique qualities: Orlando Jacob's Creek 1988 is teamed with asparagus salad, a Crozes Hermitage with steak in red wine sauce, and Château de Berbec with summer pudding. Each section is illustrated with colour photographs of the wine and finished dish, and a map showing the wine's region of origin.

Packed with useful and accessible information for the wine enthusiast and cook alike, *The Beginner's Guide to Food and Wine* is an ideal introduction for those keen to develop their knowledge of wine and a valuable aid to menu-planning for cooks and diners.

SUNNY STARTERS

PINOT OLTREPÒ PAVESE 1988

These favourite starters from the eastern Mediterranean are not easy to match

HUMMUS AND TARAMASALATA have become so much part of the foods we have around us every day that it is quite possible to forget their Near East origins. They are sometimes used for a light snack in their own right or more often for dips or starters, mostly with pitta bread, whether as part of a Greek meal or not. Yet because they are thought of as Greek foods, even though they are just as much Turkish, it is natural to assume that they would go with Greek wines. And Greek wines, low in acidity and lightly or strongly resinated as most of them are, are the complete antithesis of the lightness, crispness and freshness that marks out the vast majority of white wines on wine shops' shelves. Neither are Greek wines available as easily as the wines of other countries.

A flavour challenge

Will hummus and taramasalata, though, go only with Greek wine – or the nearest alternative available, or are there other styles of wines which will provide matches just as promising? There *are* other wines, thank goodness, that are good partners, but not all that many. One difficulty is that taramasalata and hummus have markedly different tastes from each other. Taramasalata, good taramasalata that is, not the vivid pink stuff that seems to be predominantly breadcrumbs and colouring matter, is quite oily and strongly fishy, made with smoked cod's roe (or a mixture of smoked and non-smoked cod's roe). Hummus is very different, with the more alkaline, earthy flavours and pasty textures of the pulses and a lasting strength and pungency of flavour that can kill a wine's delicate fruit. If choosing a wine for taramasalata or hummus, far more will be acceptable if you just put a little on some bread, toast or pitta bread, which absorbs some of the flavour and oil. If, though, you shovel on large quantities, you need to be more selective. Taramasalata is easier to pair than hummus. If you have both of them, either be prepared to experiment or go for one of the following suggestions.

BEST BUYS

Pinot Oltrepó Pavese 1988
Taste guide: dry

Retsina of Attica Kourtaki
Taste guide: dry

Bianco di Custoza 1988 Castelnuovo
Taste guide: dry

Côtes du Roussillon 1988 Armand de Villeneuve
Taste guide: dry

Añares 1987 Rioja
Taste guide: dry

Going Greek

One suggestion that can't be bettered is to play safe and go for Retsina, but a good Retsina and one which is lightly resinated – strongly resinated wines don't taste anything like as good in dull, grey Britain as they do in a Greek taverna in high summer! The one to look out for is

Retsina of Attica Kourtaki. Although the resinous flavour is unmistakable (particularly when the wine is sniffed), it is well balanced by a lemony character and the wine is mid-weight and not too overpowering in flavour. It sticks in the glass, though, so if you drink anything different afterwards be sure to wash the glasses thoroughly or get fresh ones. It is not overwhelmed by the flavours of either taramasalata or hummus, nor changed by them; the flavour of the taramasalata seems a little richer and that of the hummus a little more forceful, and both are every bit as good for the changes.

Italian partners

For the frequent occasions when Retsina would not fit the bill, go for the almost ideal match - from quite a surprising source. **Pinot Oltrepò Pavese 1988** is from north central Italy, just south of Milan, in a hot-by-summer cold-by-winter climate. Pale in colour it has a beautifully clean, fruity, silky and minerally aroma with a fresh, lively, zingily attractive taste and plenty of salty, aromatic fruit. Delicious on its own, it tastes a little drier, but more restrained and classier with hummus while, with taramasalata, it tastes broader and gains a lemony hint. Therefore this fresh and fruity, versatile white wine somehow manages to take on the style of the perfect hummus partner with hummus and the perfect taramasalata partner with taramasalata.

Also from Italy, but a little further west, is **Bianco di Custoza 1988 Castelnuova**. Bianco di Custoza has similarities with Soave. It comes from the southern edge of Lake Garda, not far from the Soave zone, and it uses the Garganega grape which predominates in Soave. But Bianco di Custoza also uses other grapes, mostly types which give plentiful character to a wine, which Soave lacks, so it is a bit like a super-Soave. The Castelnuovo, in particular, is a great example. Its bouquet is of crushed almonds and lemon and it tastes fresh, zesty, fruit-packed and crisp. It is a particularly good match for hummus as it provides a perfect contrast, cleaning the mouth and refreshing against the pulpiness of the chickpeas. With taramasalata, it seems slightly drier and less fruity, allowing the flavours of the fish to shine out over it.

Wines matured in oak can give some of the roundness and fleshiness of Retsina without any of its other characteristics and therefore can be good substitutes for it. If matching hummus is your primary aim, then try **Côtes du Roussillon 1988 Arnaud de Villeneuve**. It is matured in new oak barrels and the oakiness is fairly pronounced. With hummus, though, the oak is far less noticeable, allowing the wine's other flavours to shine through. It also makes the hummus taste excellent (and is quite good with taramasalata). If a taramasalata match is most important, go for white **Añares 1987 Rioja** from north east Spain. Rather broad on its own, it becomes lighter and more refreshing with really good, home-made taramasalata, which can be mixed in minutes in a food processor: process 100g/4oz skinless, smoked cod's roe to a paste with 3-4 slices of stale white crustless bread (soaked in water, then squeezed out) and 2tbls grated onion. With the motor running, gradually add 175ml/6fl oz olive oil and the juice of 1-2 lemons, blending until smooth and thick. Season with black pepper, then serve garnished with black olives and fresh parsley leaves.

PERFECT PARTNERS

Hummus

- ***Preparation: 20 minutes, plus overnight soaking***

225g/8oz chick-peas
2tbls tahini paste
juice of 2 lemons
salt and freshly ground pepper
2-3 garlic cloves
1tsp chopped parsley, 1 lemon wedge and a twist of rind to garnish

- ***Serves 6***

1 Soak the chick-peas overnight in plenty of cold water to soften the peas. (Alternatively omit steps 1 and 2 and use 1 x 425g/15oz drained can of chick-peas).

2 Wash the peas well and boil them in fresh water for 1 hour or until soft.

3 Purée the chick-peas in batches in an electric blender or a food processor with the tahini paste, lemon juice, salt and freshly ground black pepper and garlic cloves. Add enough of the cooking water and olive oil to form a thick, slightly rough cream.

4 Taste the purée, add more salt and pepper and lemon juice, if necessary, and stir in some of the cooking water if the purée needs thinning. Garnish with the parsley, lemon wedge and rind, and dribble on a little extra olive oil. Serve with warm pitta bread and taramasalata (see above).

APPETIZING APERITIFS

MANZANILLA

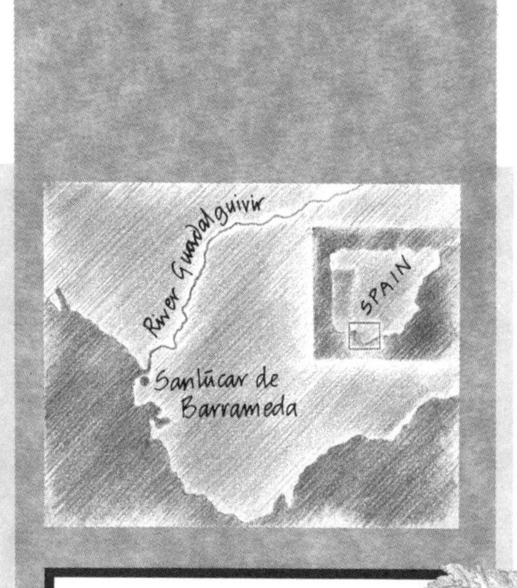

BEST BUYS

Manzanilla La Gitana
***Taste guide:* very dry**

Tio Pepe Fino Sherry
***Taste guide:* very dry**

La Ina Fino Sherry
***Taste guide:* very dry**

Fino Sherry Pale Dry
***Taste guide:* very dry**

White Port
***Taste guide:* medium dry**

Vermouth Extra Dry
***Taste guide:* medium dry**

Lillet
***Taste guide:* medium dry**

There is something safe and comforting about sticking to your favourite aperitif — but why not try out a new one for that 'pre-dinner lift'?

HOWEVER ADVENTUROUS WITH wine you and your friends are or are not, it is a good bet that you will have your favourite aperitif and stick to it whether it is G and T, white wine, Campari, Martini, a glass of orange juice or anything else. It is relaxing and pleasantly reassuring – and you know you are going to enjoy it. But once in a while, it is worth ringing the changes, either just a little, by buying a different brand of a similar drink – another vermouth, say – or, more radically, by switching to a different type of aperitif. You may well find that the loss of familiarity is easily compensated for by the enlivening effect of a taste you actually notice.

The most classic and, sadly, most underrated and misunderstood aperitif is sherry. Real sherry comes only from Spain, from its hot, dry south-west corner, and its production is concentrated around the town of Jerez (hence the name) in Andalucia. For the most palate-awakening aperitif anywhere, look out for one called fino. It is light in colour, bone dry, with a distinctive fresh, floral, salty and curdy flavour that comes from its remarkable method of production. After the grapes are picked, fermented and fortified, the wine is put into butts – smallish oak barrels – to mature. Then, a rare yeast type, called flor, which occurs naturally in the atmosphere, is attracted to the wine and grows on its surface, forming a thick, creamy white layer. The layer not only protects the wine from the harmful oxygen in the air, but the yeasts react with the wine and gradually change its character for the better, turning it into fino sherry.

Freshness first

Once bottled, fino sherry loses its protective covering and, never having been acclimatized to oxygen, it is very susceptible to its damaging effects. This means that it is very important to buy it as fresh as possible. But since there is no date on the label it can be difficult to judge just how old it is. To avoid disappointment, buy one of the major brands, which sell fast, from a busy shop or supermarket. **Tio Pepe** and **La Ina** are both top-quality fino sherries which are usually in perfect condition and have impeccable fino character. Or, for a bottle that is very nearly as good, but considerably cheaper, go for **Fino Sherry Pale Dry** bottled by E Palomo & Co.

Sherry that comes from the little town of Sanlúcar de Barrameda, near to Jerez but on the coast, matures in a slightly gentler climate which is even more beneficial to the growth of flor. The sherries are therefore even lighter and more flowery and, instead of fino, are called manzanilla. The most elegant and delicious example you can find is **Manzanilla La Gitana** from Vinicola Hidalgo. It is 2% lower in alcohol than most sherry, too, which gives it an even more delicate and appetite-tempting flavour.

Keep these dry sherries in the fridge once opened, and serve them chilled. Nothing goes better with them than two plates of olives: one green, one black. Or, for more adventurous nibbles, you could make a whole range of tapas such as tortilla, prawns or chicken livers.

Port to heaven

You probably think of port as something for after the meal, not before it. So it is, usually. But there's also **White Port.** It smells, well, not dissimilar from red port really. Its taste is also similar but much lighter and less sweet; it makes a surprisingly refined aperitif. It should also be served well chilled, or over ice – which will help dilute its rather strong 20% alcohol. The best accompaniment is nuts.

Variation on a theme

If you normally drink Martini (without the gin) or Cinzano and fancy a variation on the theme, there's **Vermouth Extra Dry.** It has the most extraordinary flavour of oranges, limes, peaches, mangoes – a whole fruit salad with herbs macerated in it, which can't fail to make you sit up and take notice. It will balance the most powerfully flavoured nibbles, or baby pizza pieces, vol-au-vents, and so on. It is also a tremendous bargain. Serve it with lots of ice.

For a complete change, try **Lillet.** It was once very popular here, then fell out of fashion and completely disappeared. Now it has been revived, and you can see why it was so much liked before. It is actually produced in Bordeaux and is a mixture of white Bordeaux wines and fruit-based liquors, aged in oak. It smells and tastes like a cross between a wine and a liqueur, and is very orangey - the predominant flavour of this delightful aperitif is that of orange peel.

This light and tangy, citrus-flavoured aperitif can be drunk either on its own, or over ice. In the summertime, you might like to fill a tall glass with ice cubes, pour over the Lillet, then add a splash of soda water or sparkling mineral water and decorate with a slice or two of fresh orange. Its accompaniments are best simply flavoured, or cheesy, so they don't spoil the orange tastes - a wedge of tortilla (see below) is perfect.

PERFECT PARTNERS

Tortilla

The custom of nibbling *tapas,* such as tortilla, with an aperitif is widespread in Spanish bars

- ***Preparation: 15 minutes***
- ***Cooking: 20 minutes***

3-4 eggs
5-6tbls olive oil
1 large potato, peeled and cut into cubes
2tbls finely chopped onion
salt and pepper
knob of butter, to glaze

- ***Serves 4***

1 Break the eggs into a bowl, add 1tbls water and stir them vigorously with a fork or a wire whisk.

2 Heat 3tbls of the olive oil in a small frying pan. Add the diced potato and finely chopped onion and cook for about 10 minutes, until the potatoes are soft, stirring from time to time. Remove from the pan with a slotted spoon, add to the eggs and season to taste with salt and pepper.

3 Heat a 15cm/6in heavy-based frying pan over a moderate heat, then add 2tbls olive oil. Cook the tortilla, lifting the edges to permit the liquid egg to run underneath and shaking the pan to prevent sticking. Continue until a golden crust has formed underneath.

4 Turn the tortilla onto a plate, scrape off any bits adhering to the pan, then heat a little more olive oil in the pan – about 1tsp. Slide the tortilla back into the pan to brown the other side. The surface of the tortilla should be golden and slightly crisp, the inside almost runny.

5 Slip the tortilla on to a serving plate and glaze with butter. Serve hot or cold, cut into wedges.

EASY AVOCADOS

PINOT BLANC EGUISHEIM

Thirty years ago, the avocado was a rarity in Britain, but now it is a favourite starter

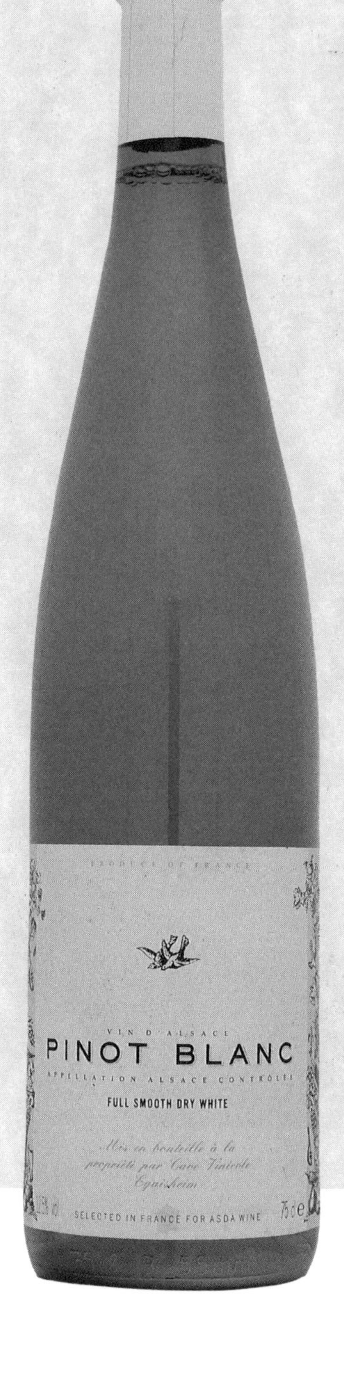

THE AVOCADO IS versatile: it can be simply halved and filled with a dressing or a simple salad, or made into a mousse or a spicy dip. It is quick and easy to prepare and has a fine flavour when in perfect condition, never too powerful, acidic, peppery or bitter – in short, it has none of the tastes that should make it a difficult match for wine. Yet a difficult match it is, for although its flavour is subtle, it has a texture that stays in the mouth for a long time and can take away some of the fruit flavours in wine.

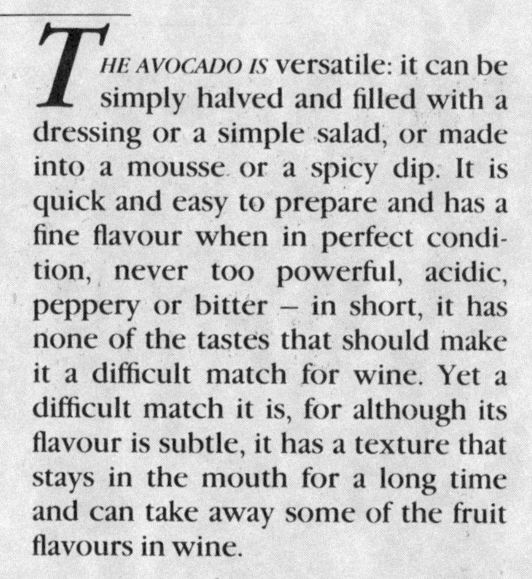

A wine of character

The solution would seem to be a wine that bursts with fruitiness so that even if this is diminished when it accompanies avocado, there will still be enough left to be appreciated. Alas, many fruit-packed wines are just too much for the delicate avocado flesh: the wine will taste fine, but the avocado will be overwhelmed. Ideally, the wine should be good enough to accompany the rest of the meal. A choice that is hard to beat is **Pinot Blanc Alsace** from the Cave Vinicole Eguisheim. Alsace, on the French side of the Rhine, just east of the Vosges mountains, is one of France's sunniest spots and despite the region's proximity to the German wine districts which accounts for the aromatic fruitiness of the wine the structure of Alsace wines is French. Pinot Blanc is an excellent choice because the grape variety gives just the right amount of lively fruit bouquet and taste. It is one of the few combinations of avocado and wine where both, when consumed together, taste just as good as on their own.

Magic Moselle

For a wine with more pronounced fruitiness, go for one from the German side of the border: **Mosel-Saar-Ruwer Kabinette 1985**. Mosel-Saar-Ruwer is the region the wine comes from, an area of steep, twisting, slatey terraced vineyards in the valley of the Mosel (Moselle) and its tributaries, the Saar and the Ruwer. The wine actually comes from the Bereich (District) of Bernkastel and has a floral, blackcurrant aroma and is full of rounded fruitiness. It has only 8% of alcohol, an extra benefit!

BEST BUYS

Alsace Pinot Blanc Eguisheim
Taste guide: dry

Mosel-Saar-Ruwer Kabinett 1985
Taste guide: medium dry

Touraine Sauvignon 1988 Domaine Joël Delaunay
Taste guide: very dry

Faustino V 1988 Rioja Blanco
Taste guide: dry

Zagarron 1988
Taste guide: very dry

Vin de Pays des Côtes de Gascogne 1988
Taste guide: dry

Something a little drier

If you need a wine to partner the whole meal and your main course is fish, or if your avocado has a seafood filling, **Touraine Sauvignon 1988 Domaine Joël Delaunay** may well be the one to go for. It is from the Loire valley of north-central France and epitomises the crisp nettles-and-gooseberries taste of the Sauvignon grape. Young and therefore at its best, it has plenty of fruit but its sharpness stops the fruit being too overwhelming and makes it a good avocado partner.

Support from Spain

For extra body and a wine that will support a variety of different dishes, Northern Spain provides the answer with **Faustino V 1988 Rioja Blanco**. Its alluring bouquet gives the sensation of a rich, rounded wine and this is confirmed on the taste, though the wine is also dry and firm. It is not too fully flavoured for the avocado, though when drunk against it, Faustino V becomes weightier and firmer. Also from Spain is the light, crisp greengage-and-mango flavoured **Zagarron 1988**. It comes from the Airen grape and from Spain's centre, the vast upland plain of La Mancha where the image of Don Quixote and his windmills still pervades the atmosphere. The wine loses a bit of its fruit and seems a bit steelier when drunk with avocado, but the marriage of flavours is still very good. For one of the most quaffable, versatile, good value wines around, go no further than **Vin de Pays des Côtes de Gascogne 1988.** Its crispness and its assertive appley fruit don't change the flavour of avocado and the wine itself stays dry and firm.

Whatever your choice of wine, make sure that the avocado is ripe. Buy still fairly firm, at least two or three days before you need it, and wrap in a brown paper bag. Store the avocado in a not-too-cool, dark place - a kitchen drawer is ideal - until it will just give to gentle pressure from your thumb. Cut in half, then brush the cut surface with a little lemon juice.

PERFECT PARTNERS

Avocado starters

One of the simplest and most delicious starters in the world is a perfect avocado, halved and served with a vinaigrette dressing, or even an avocado simply sprinkled with lemon juice, salt and pepper. Or add soured cream, yoghurt or crumbled Roquefort cheese to the vinaigrette for a refreshing flavour. In Mexico avocados are puréed and blended with sour cream, lemon juice and spices to make guacamole – eat with tortilla chips or a selection of crisp, raw vegetables – or use the dip as a filling for hard-boiled eggs. Mix with cottage cheese to make another simple, tasty dip (see Perfect Partners page 20).

Avocados have a natural affinity with seafood and are delicious filled with peeled prawns dressed with mayonnaise that has been flavoured with a little tomato ketchup and a dash of hot pepper sauce. You can dress up tuna fish, flaked crab, poached white fish or even cold cooked chicken with the same sauce: leave the flesh in the shell, or take it out, dice it and add to the filling, then heap the mixture back into the empty shells. Or slice the avocado thinly and serve with a heap of dressed crab.

For a quick and unusual snack, pile mashed avocado on toast with a dash of lemon juice, a smidgeon of pepper sauce and a sprinkling of toasted sesame seeds. A glass of your chosen wine – delicious!

CELEBRATE WITH CHAMPAGNE

BRUT RESERVE

Open a bottle of bubbly and celebrate in style – for special occasions there is no substitute for Champagne

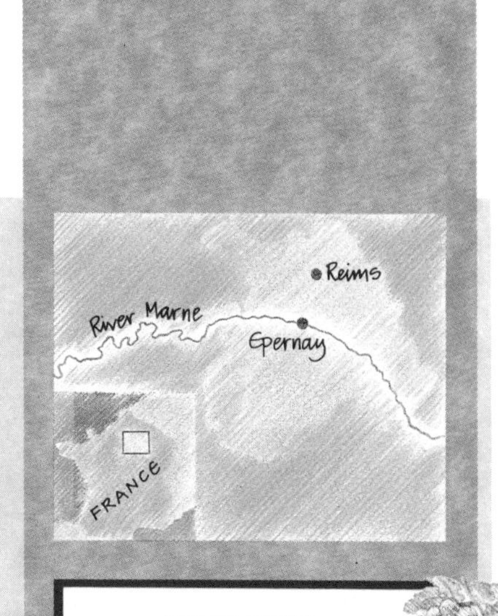

BEST BUYS

Brut Reserve Special Cuvée Epernay
***Taste guide:* dry**

St. Michael Premier Cru Brut
***Taste guide:* dry**

Lambert Cuvée Exceptionelle 1983 Brut
***Taste guide:* dry**

Louis Bernard Brut Rosé
***Taste guide:* dry**

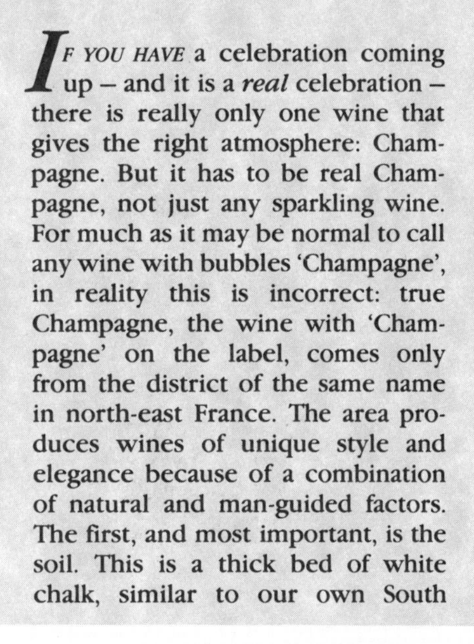

IF YOU HAVE a celebration coming up – and it is a *real* celebration – there is really only one wine that gives the right atmosphere: Champagne. But it has to be real Champagne, not just any sparkling wine. For much as it may be normal to call any wine with bubbles 'Champagne', in reality this is incorrect: true Champagne, the wine with 'Champagne' on the label, comes only from the district of the same name in north-east France. The area produces wines of unique style and elegance because of a combination of natural and man-guided factors. The first, and most important, is the soil. This is a thick bed of white chalk, similar to our own South Downs and the white cliffs of Dover, which retains an even temperature whatever the season.

The particular climate, which

STORING CHAMPAGNE

If you have a special occasion like a birthday, an anniversary or, particularly, a wedding coming up, buying the Champagne is probably not one of the things you think of doing first. But it should be, because buying Champagne just a few months before you need it and storing it carefully can improve it tremendously and make it taste like a bottle costing many pounds more. Champagne needs time to mature; it improves immeasurably in bottle. But sadly there is so much demand for the wine that the moment it is released by the cellars, it is shipped, on the shops' shelves and sold. If you buy a bottle, especially one of the more affordable Champagnes from supermarkets, and drink it straightaway you may find it tastes disappointingly harsh and green. Keep a bottle three (or better six) months and then drink it and it will be a revelation.

Ideally, therefore, you should aim always to keep a bottle or two in stock, in preparation for the next celebration, impromptu or otherwise. Try to find a spot well hidden away (to stop you being tempted to drink them sooner), somewhere not subject to huge temperature variations or vibrations, and keep the bottles lying flat to stop the corks drying and shrinking and the bubbles escaping.

only just about permits the grapes to ripen properly – some years hardly at all, occasionally abundantly – is important, too. For it is reckoned that when the vine has to struggle to produce its grapes, those grapes turn into particularly refined wine.

Grape-picking

Then the choice of grape varieties has its part to play. Surprisingly, Champagne is usually produced from Pinot Noir, Pinot Meunier and Chardonnay, ie a mixture of black and white grapes, with the majority black. This means that as soon as the grapes are picked the juice (which is always white) has to be pressed very quickly, before any colour can leach from the dark skins.

Sometimes only the white Chardonnay grapes are used. Then the wine is called 'Blanc de Blancs', a short way of saying white (blanc) wine from white (blancs) grapes. One of the nicest, and most affordable, of these is **St Michael Premier Cru Brut** bottled by Union Champagne Avize. It has very small, even bubbles – a sure sign of quality – and a delicious yeasty, biscuity bouquet, which is the hallmark of carefully made Champagne.

If you prefer the slightly fuller and more typical style of Champagne made from black grapes as well as white, look for **Brut Reserve Special Cuvée Epernay.** The colour is an attractive light gold, the bubbles keep rising. The bouquet is gentle, yeasty and a little appley, while the taste is dry and mellow.

Champagne is usually non-vintage. This means that although most of the wine in the bottle will come predominantly from the grapes of one harvest, producers are able to blend in small quantities of older wines, which they keep just for the purpose.

A vintage crop

Sometimes, though, about once every three years on average, the harvest looks particularly promising and producers then select a proportion of the crop to make a vintage wine, that is, a wine solely from that year's produce. If your celebration is worth the cachet of vintage, go for **Lambert Cuvée Exceptionelle 1983 Brut.** Bright gold, with tiny, long-lived bubbles, it has an unusual but very attractive yeasty bouquet. Its taste is mature and opulent with enlivening, balancing acidity.

Or there is pink Champagne, once the height of fashion, where a little colour from those black grapes *is* permitted to tint the wine. For a pretty, light rosé hue and a gentle floral, fruity style, you can't fail to enjoy **Louis Bernard Brut Rosé** – it looks and tastes as good as the occasion makes you feel.

Party puffs (below) are a perfect nibble to serve with Champagne. You will need to make a savoury choux paste: place 100g/4oz butter in a pan with 300ml/½pt cold water. Bring to the boil, then remove from the heat, tip in 150g/5oz seasoned flour and beat until thick. Beat in 4 eggs a little at a time.

PERFECT PARTNERS

Party puffs

- ***Preparation: making choux paste, then 20 minutes***
- ***Cooking: 25 minutes, plus cooling***

butter, for greasing
75g/3oz Gruyère cheese, grated
1 quantity savoury choux paste (see above)
1 beaten egg, for glazing
For the fillings:
225g/8oz full-fat soft cheese
4tbls double cream
4tbls finely snipped chives
pepper
225g/8oz smoked salmon trimmings
125ml/4fl oz soured cream
2tsp lemon juice
sprigs of parsley, to garnish

- ***Makes 36 puffs***

1 Heat the oven to 200C/400F/gas 6. Butter two baking sheets. Beat the cheese into the choux paste mixture while it is still warm. Fill a piping bag with a medium-sized plain nozzle with the choux pastry. Pipe out small bun shapes, about 1 tablespoonful in size, onto the buttered sheets, 2.5cm/1in apart. Brush the puffs with the beaten egg to glaze.

2 Bake the choux puffs for 15 minutes, then reduce the heat to 190C/375F/gas 5 and cook for a further 10 minutes. Remove from the oven and make small slits in the puffs to allow the steam to escape. Cool on a wire rack.

3 Meanwhile, make the fillings. Beat the cheese until light and fluffy, then beat in the double cream. Add the chives, season with pepper and set aside.

4 Pound the salmon trimmings to a smooth paste, stir in the soured cream and lemon juice and season with pepper.

5 Shortly before serving, split open the cold puffs and fill half of them with the chive filling and half with the smoked salmon filling. Garnish with sprigs of parsley.

WEDDING FIZZ

CREMANT DE BOURGOGNE

The big day is dawning, the guests have been invited, the food has been planned. Now you must choose a drink to toast the happy couple

CATERING FOR THE fizz for a wedding makes the most of your powers of compromise. Of course you want to give your guests the best on such a special occasion, and there's no doubt there's nothing nicer, nor more celebratory, than real champagne. But champagne is costly, and unless everyone is to be rationed to just one glass for the toasts, you will need to cater for half a bottle per head – more if it is purely a champagne reception. Even if it is to be restricted to one glass each, you will need some other sparkling wine for more liberal drinking.

On the other hand, it is pointless buying a hangover-inducing sparkling wine from the bargain basement that no one really enjoys and which gives your guests more the impression that they are at a casual gathering than helping you to celebrate what should be the most important day in your life.

BEST BUYS

Crémant de Bourgogne 1985
Taste guide: dry

Crémant de Bourgogne 1985 Les Caves de Bailly
Taste guide: dry

Gratien & Meyer Brut, white and rosé
Taste guide: dry

Segura Viudas
Taste guide: dry

'METHODE CHAMPENOISE'

On the labels of most better-quality sparkling wines you will see the words *méthode champenoise*. This means that the wine has been made sparkling by the best, but unfortunately most expensive, method. This involves a long second fermentation in bottle to create the fizz, then many days twisting and tilting the bottle to work the resultant sediment down to the neck end before whipping out the cork and its sediment plug, righting the bottle and recorking it with fresh wine to make up the shortfall.

The EC has decided, though, that in order to avoid any confusion with real champagne the term *méthode champenoise* will have to be phased out. Different countries and producers are, therefore, developing alternative terms to describe their best sparklers. For example, you will soon be seeing *méthode traditionelle* from France and *metodo classico* from Italy. Spain, on the other hand, has no problems. Traditionally it has always preferred to stick to its own description, *Cava*.

While there is no such thing as a real substitute for champagne – its stylish flavour is unique – it is possible to find a number of sparkling wines which will stand in its place.

Stylish sparkles

One of the best 'champagne substitutes' is Crémant de Bourgogne. Its main link with champagne is that it is made from the Chardonnay grape variety, which is an important champagne constituent. It all comes from the Burgundy region of eastern France, much from the exemplary Cave de Viré in the district that also produces Macon. Crémant actually means 'creaming' and refers to a sparkling wine with a little less than its full quota of fizz: the effect in the mouth is less aggressive and creamier. But there's still plenty of long-lasting sparkle – in fact Crémant de Bourgogne can be very lively, so open bottles with care! **Crémant de Bourgogne 1985**, labelled Brut, which means dry, is dry indeed, but not at all harsh. It is well-rounded, soft, biscuity and with a salty tang – your guests will be impressed by its quality.

Pretty in pink

Or, for something a little different, why not go for the pink version of **Crémant de Bourgogne 1985** from Les Caves de Bailly, made predominantly from Pinot Noir, which is another important champagne grape variety? A festive pale rose colour, it has good raspberry-like fruit superimposed on the biscuity saltiness and is a little fuller and richer - perfect for both summer and winter weddings.

Another area that is often put forward as a good hunting ground for non-champagne sparklers is the Loire valley in north-west France and, in particular, the part of it around Saumur. This is the territory of the Chenin Blanc grape variety, one that is rarely seen in other regions but which has the great advantage of being equally suitable for dry, sweet and sparkling wines. A good example of sparkling Saumur is **Gratien & Meyer Brut.** A characteristically light colour, it has a lemony fruit salad of a nose. Its taste is light but flavoursome with freshness and fruit, rather like the wine equivalent of a sweet and sour dish. It too has a rosé counterpart which has just a little extra body and fruit.

Traditional tastes

Spain is another country making excellent sparkling wines. Most of the best come from the north east, just behind Barcelona, in the region known as Penedés. Made from local grape varieties, at their most traditional the wines can be surprisingly big and earthy. **Segura Viudas** though, combines traditional methods with modern style to turn out a full, clean well-balanced, characterful wine of class - robust enough to serve with a flavourful, cheesy party sandwich loaf (below).

PERFECT PARTNERS

Party sandwich loaf

• ***Preparation: 1¼ hours, plus 2 hours chilling***

1 large, 1-day-old white loaf
300ml/½pt mayonnaise
1 bunch of watercress, chopped
3 tomatoes, skinned and sliced
2 eggs, hard-boiled and sliced
1tbls tomato ketchup
pinch of cayenne pepper
dash of Worcestershire sauce
150g/5oz canned white crabmeat (drained weight), finely shredded
8 lettuce leaves, shredded
For the cheese coating:
250g/9oz full-fat soft cheese
salt
juice of ½ lemon
pinch of paprika
3-6tbls double cream

• ***Serves 8-10***

1 Trim the crusts from the loaf and cut it lengthways into five even slices.

2 Make a watercress filling by mixing half the mayonnaise with the chopped watercress and spread the first slice of bread with half of it. Top with half the tomatoes and half the eggs, then the second slice of bread.

3 To make the crab filling, blend the rest of the mayonnaise with the ketchup and season with cayenne and Worcestershire sauce. Mix in the crabmeat and shredded lettuce. Spread the second and third slices of bread with this filling.

4 Spread the fourth slice of bread with the remaining watercress filling and cover with the remaining tomatoes and egg.

5 Finish the sandwich with a plain slice of bread, then wrap the loaf in foil and chill for 2 hours or until firm.

6 Meanwhile, prepare the cheese coating: in a large bowl combine all the ingredients, beating until smooth. Spread over the top and sides of the loaf.

PARTY REDS

PAVLIKENI

There can be no more excuses for drinking vinegar at parties when there are so many respectable, reasonably priced and readily available red wines about

BEST BUYS

Pavlikeni, Bulgarian Country Wine
***Taste guide:* light-bodied**

Bellerive
***Taste guide:* light-bodied**

Vin de Pays des Côtes Catalanes
***Taste guide:* medium-bodied**

Claret
***Taste guide:* medium-bodied**

Sicilian Red Wine
***Taste guide:* medium-bodied**

SOMEHOW IT SEEMS much easier to find enjoyable, easy-to-drink red wines at bargain prices than it does white. In fact, even if you generally prefer white wine, at a party you might do well to try the red instead. Similarly, if you have to pop into an unfamiliar shop at the last minute for a bring-a-bottle do, especially one where there's a chance that the wines might be discussed, or at least commented on, you'd be wise to look at the reds. Good bets include wines labelled Vin de Pays from France, or wines from Italy, Portugal or Bulgaria. If, though, you remember to get the wine with your normal shopping in the days before the party, or if you want to stock up for spur-of-the-moment invitations, you can find a large selection of really good bottles that even a wine buff wouldn't turn his or her nose up at.

Although wine with no denomination

COOKING WITH WINE

If you like keeping left-over wine for cooking, don't hang on to it for more than two or three days before using it. Wine, once opened, can go vinegary or oxidized which makes it flat, dull and unpleasant. When you cook with wine its flavours are concentrated, so if you use bad wine, you'll get concentrated bad flavour.

By far the best wine to use is a little of the one you are planning to drink with the meal you are preparing. You don't need a lot and it will ensure the dish and the wine go well together. But if you can't bear to throw away wine dregs, keep them tightly closed with as little air space as possible – and plan your recipes to use them up as soon as you can.

Once wine starts cooking, the alcohol boils off (which may or may not be a good thing, depending on your viewpoint) as will several flavour elements and all the aroma, so if you want *real* wine taste it has to be added at the last possible moment. Otherwise, the essence of cooking with wine is to do it as gently as possible so its flavours get absorbed into the food, not boiled away.

other than simply Vin de Table and a brand name can be incredibly basic, sometimes it can be a real winner. One such is **Bellerive.** Its colour isn't too deep, indicating fairly light body, but it has a delicious soft, savoury taste and a velvety consistency that would go down a treat either as a quick glass with the neighbours or throughout an evening. A pleasantly dry finish means it doesn't cloy and would go well with all those party nibbles that you hope will be circulating.

Nowadays wine stores are putting more and more effort into choosing their house wine. One inspired selection is **Vin de Pays des Côtes Catalanes.** It comes from Mediterranean France, on the border with Spain, and seems to have acquired the good points of the wines of both these countries. A bright, cheerful-looking purplish red, it has plenty of flavour, good acidity giving it structure, and a ripeness that comes from its southern origins, making it a really characterful drink. It would be at its best at parties where more substantial food is served but it can be drunk very happily on its own.

Casual Claret

Another wine that would be ideal for parties with 'real' food rather than just nibbles is **Claret.** Now claret means no more or no less than any red wine from Bordeaux and it comes in hundreds of different styles, qualities and prices. The one to look for is bottled by Yvon Mau. Quite deep in colour, it has a cedary, blackcurranty nose, succulent wine gum and liquorice flavour and good fruit, but it isn't too fully flavoured or too tannic, so is just the right weight for casual drinking.

Popular Pavlikeni

For something that can't fail to be popular with guests if you are the host of a party, yet is a just a little bit different, try **Pavlikeni, Bulgarian Country Wine.** This wine is a clever blend of the Cabernet Sauvignon grape variety, one of the most successful red grapes ever produced, with the softer and fruitier Merlot. It comes from a particular part of Bulgaria: Pavlikeni. 'Country Wine' is the Bulgarian equivalent of French 'Vin de Pays'. It is dry, yet soft and attractively plummy, and so well rounded that it slips down almost without your noticing, leaving a warming spicy taste in the mouth.

If what you want above all else is fruit, another modestly named wine gives you bags of it. **Sicilian Red Wine,** a humble Vino da Tavola, bottled by Società Industriale Vinicola SpA is a great combination of the deeply coloured, sweet, ripe grapes from Sicily's south east tip and the Barbera grape, transplanted from its homeland in north west Italy. The result is all the fruitiness of a perfect summer pudding, with cinnamon and clove flavours and a savoury tang. The taste leaps out of the glass with such vibrancy that it would be a big hit at any party, as would the marinated green and black party olives (see below).

PERFECT PARTNERS

Party olives

These olives will keep for a long time in screw-top jars. The oils can be used in salad dressings

- ***Preparation: 30 minutes, plus marinating***

For Lemon and coriander olives:
450g/1lb black or green olives, drained if in brine
1tbls coriander seeds
1 lemon, thinly sliced
olive oil, to cover
For Garlic olives:
450g/1lb black or green olives, drained if in brine
6 large garlic cloves, crushed
olive oil, to cover

- ***Makes 900g/2lb***

1 Discard any bruised olives. With a thin, sharp knife cut two or three shallow slits in each olive, wash well under cold running water and drain.

2 To make the Lemon and coriander olives, mix the drained olives with the coriander seeds in a large bowl. Pack the mixture into a large, clean jar, adding thin slices of lemon here and there as you proceed.

3 Pour olive oil into the jar to cover the olives and lemon slices. Seal the jar and keep in a cool place for at least a week to mature.

4 For the Garlic olives, place the olives and the crushed garlic in a large, clean jar and cover them with olive oil. Allow the olives to marinate for at least 3 days.

5 To serve, drain the olives, discarding the lemon slices from the Lemon olives.

SAFE WINES TO TAKE ALONG

CHARDONNAY D'OC

BEST BUYS

Chardonnay 1988 Vin de Pays d'Oc
***Taste guide:* Very dry**

Seaview 1988 South Australian Chardonnay
***Taste guide:* Dry**

Chablis 1986
***Taste guide:* Very dry**

Montana 1987 Sauvignon Blanc/Chenin Blanc
***Taste guide:* Medium dry**

Vernaccia di San Gimignano 1987 San Quirico
***Taste guide:* Dry**

What wine can you take with you to a dinner party when you don't know what's on the menu or what is the host's choice of drinks?

IF YOU ARE going to a dinner party and decide to take a bottle of white wine with you, do chill it well before leaving home. Unless you are going on foot and it's really cold outdoors the wine is sure to warm up en route and your host may not have the time or fridge space to cool it for the first course. Similarly – unless you're sure you'll be first to arrive and your host won't have anything already opened – don't take anything along that's specifically an aperitif.

Safe bet

So, what to take that no one will dislike, is versatile enough to go with whatever will be served but can be happily drunk on its own as an aperitif? No wine will please everyone every time but several go with a wide variety of meals and please many tastes.

Safest bet is something made from Chardonnay grapes. These grapes have now been planted so widely throughout the world because their creamy-buttery flavour lends itself so well to so many different wines.

Lemony Chardonnay

A clever choice would be **Chardonnay 1988 Vin de Pays d'Oc,** from the south of France, not far from either the Mediterranean or the Pyrenees, where it benefits from the cool ambience of the Mediterranean and the Atlantic. It is also vinified very carefully, fermentation temperatures being kept low to preserve the fruitiness. This wine, light in colour, smells lemony-crisp and just slightly cedary, as 20 per cent of it is made in oak casks. Its taste is remarkaby fresh, but with a good richness and a rounded slightly toasted edge. Few will not be charmed by it and few dishes will not be complemented by it.

Restrained partners

From the Antipodes comes **Seaview 1988 South Australia Chardonnay.** Australian wines are now very popular – and rightly so. The Australians have been developing their wine industry in the most determined way, making finer and finer wines. The Seaview estate does not produce those really weighty and concentrated Australian wines. Though their products are typically Australian, and typical of the variety, they are restrained enough to partner most foods, not outweigh them.

Seaview Chardonnay has a big, buttery-oily bouquet and a full, herby - lemon - creamy - pineapple taste that carries hints of other exotic fruits. It is versatile enough to go with all but the most delicate of starters and will easily win over your fellow guests.

Ageing gracefully

If you feel that your hosts are a little conventional in their tastes and their cooking, and if you don't mind spending a little extra, then **Chablis 1986** is for you. Most Chablis to be seen about is quite young and can be drunk young. But Chablis can age very well, and the '86 is now mellow, with a rather elegant lemony-buttery nose and a well balanced taste. A steely acidity balances soft rounded fruit. It cannot be bettered as a partner for a fish course or seafood starter.

Dry but flowery

Deciding on how dry or sweet a wine to take with you can be quite a problem, especially if you don't know your fellow guests well. Some find dry wine sour; others dislike anything that's not bone dry. A fair compromise is **Montana 1987 Sauvignon Blanc/Chenin Blanc.** This blend of the two grape varieties gives a dry gooseberry edge – the Sauvignon – with a slightly-sweet touch of flowers – the Chenin Blanc. Though its sweetness is minimal its ripe fruitiness will appeal to "sweet" people. It comes from the Marlborough area in the north of New Zealand's South Island.

New Zealand's benign climate is ideal for producing white wines with plentiful aroma, rounded fruit and fresh bite. This particular wine is a prime example and fellow guests would be interested to try it.

Tuscan choice

Italian wines are, of course, remarkably good food partners. an unusual but wise choice for a dinner party is **Vernaccia di San Gimignano 1987 San Quirico.** This wine, product of a hilltop Tuscan town of sun-baked medieval towers, is made from the local Vernaccia grapes and is dry, full and clean, with flavours of apples, nuts and a touch of earthiness. It is not right for an aperitif, but try it with starters of any kind – or even a main course – and see it come into its own.

Some dinner party dishes are surprisingly difficult to match with wines, so it is a good idea, if you can, to find out what is on the menu. Avocado starters such as avocado dip (see below) are particularly tricky to find a partner for. However, you will find some useful pointers and suggestions on pages 11-12.

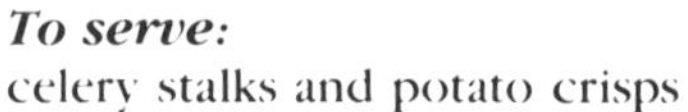

Avocado dip

• ***Preparation: 10 minutes***

1 large ripe avocado
175g/6oz cottage cheese
1 large ripe tomato, skinned and chopped
1 small onion, finely chopped
lemon or lime juice
1-2 drops hot pepper sauce
salt

To serve:
celery stalks and potato crisps

• ***Serves 6***

1 Halve the avocado and remove the stone. Scoop out the flesh and mash to a coarse consistency. Mix in the cheese, tomato and onion; season with lemon or lime, hot pepper sauce and salt to taste. (Alternatively, for a smoother consistency, purée the dip briefly in a blender or food processor.)

2 Serve the dip in a bowl surrounded with celery stalks, potato crisps and savoury nibbles for dipping.

WINES TO BRIDGE THE GAP

JURANÇON

Carefully chosen, white wine can be served as an aperitif, then carried on to accompany the starter, and sometimes a main dish as well

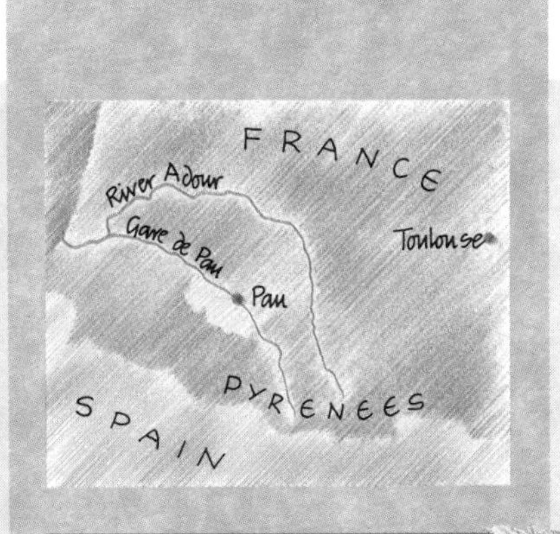

BEST BUYS

1987 Jurançon Cave de Producteurs
Taste guide: sweet

Beaujolais Blanc
Taste guide: dry

1987 Quincy Vin Noble Domaine de Maison Blanche
Taste guide: very dry

Orvieto Classico Secco 1987 Antinori
Taste guide: dry

Gewürztraminer Rheinpfalz Qualitätswein
Taste guide: dry

APPETIZING AS SHERRIES, vermouths or other aperitif drinks may be, it's often nicer (and less alcoholic!) to have a decent glass of white wine before eating. But unless there are six of you it's wasteful to open a bottle for just one glass each before going on to a different wine with your meal. The answer is to choose the sort of wine that is fine in the perking-up, appetite-enhancing role of the aperitif while also a good partner to your first course.

Light and dry

An aperitif wine shouldn't be too weighty, rich and concentrated, or else it will jade the palate and dull the appetite – the reverse of what is required. For example, if you were planning a creamy-sauced starter, particularly one based on fish, you might decide to serve a smart white Burgundy. But if you want the wine to act as aperitif as well it would be better to go for **Beaujolais Blanc**. This still has the buttery, biscuity, earthy character of the Chardonnay grape from which both white Burgundy and Beaujolais Blanc are made but, as the latter comes from further south in the same central-east part of France, it is just that important bit less heavy – so much better suited to drinking on its own. It would also 'bridge the gap' quite nicely if you were having just one course as it goes well with poultry and other fish dishes.

Crisp and fresh

Along the same lines of thought, instead of a Sancerre with seafood, oily fish, salad-based or pastry-based starters, try **1987 Quincy** (pronounced Cahn-see) **Vin Noble Domaine de Maison Blanche**. Like Sancerre, it comes from the middle part of the Loire Valley in northern France and, like Sancerre, it is made from the Sauvignon grape variety. But the area of Quincy is a little further from the Loire river, where the Sauvignon grapes don't develop their very assertive, grassy, gooseberry-like character to the same degree. Crisp and fresh, the Quincy still has a hint of gooseberries, and of green beans (!), and has good intensity of flavour – but not too much. It would also go with all white meats, vegetable-based dishes and salads.

Soft and creamy

Italians hardly ever drink their wines on their own. The wines are made, they say, to be drunk with food, and show their best only when so drunk. Yet, surprisingly enough, many of them *are* excellent as aperitifs, particularly for those whose stomachs are sensitive – especially when empty – to more acidic wines. One of the best with-and-without wines is **Orvieto Classico Secco 1987 Antinori**. The word *secco*, meaning dry, is important. Orvieto is also made *abboccato* (medium) which can be just as pleasant on its own, but is adaptable to far fewer starters. *Classico* means that the wine comes from the best part of the Orvieto district for vine cultivation, so is also worth looking out for.

Orvieto comes from around the town of the same name in the region of Umbria in central Italy. Umbria is one of the very few Italian regions with no coastline; it is full of fascinating Etruscan remains.

Its wine can be just as fascinating. The Antinori Orvieto is soft, creamy, lightly nutty, and an ideal gentle pick-me-up as well as being very versatile in the choice of first courses it would accompany. Anything from pasta in a fishy sauce to Parma ham and melon would go – and it's just as good with starters of non-Italian origin.

Slightly sweet

It is often more satisfying to have a slightly sweeter drink before meals. If wine is your choice and you want it to partner your food too, there are two solutions. One is to serve a dry wine, but one that isn't bone dry and that actually tastes sweeter than it is. One such is **Gewürztraminer Rheinpfalz Qualitätswein**. It is from Germany (even though Gewürztraminer is a grape variety seen much more often from Alsace in France) and, in particular, from the Rheinpfalz region where the wines have more body than usual. This wine is soft, with the lychee-like character of Gewürztraminer, but doesn't have its characteristic spiciness, making the wine much lighter and easier to drink. It also has a hint of asparagus on the nose, which is a huge giveaway: it goes like a charm with asparagus – but could also accompany sausages, be they salami, wurst or even just bangers.

Lightly honeyed

The other answer is to choose a sweeter wine, but not to have too sharp, too lean or too fruity a starter. The ideal choice is **1987 Jurançon** from the Cave de Producteurs. It isn't sticky sweet, just lightly honeyed and as crisp as a fresh apple; a delicate and indulgent glassful. It is excellent, too, with all the salty snacks and nibbles, such as roasted nuts or potato crisps, that may precede the meal. When it comes to food, though, the best partner for this light, mildly honeyed, crisp white wine is pâté, whether as a starter or a main course. Jurançon would shine with foie gras, by far the richest of all pâtés, yet it would also be equally good with a less rich, everyday pâté such as one made with chicken livers (see below). However, any rich, flavourful meat- or cheese-based starter, or one made with mushrooms, would be enhanced by this wine.

Chicken liver pâté

PERFECT PARTNERS

- ***Preparation: 15 minutes, plus chilling***
- ***Cooking: 15 minutes***

50g/2oz butter
4tbls finely chopped onion
225g/8oz chicken livers, trimmed
2 hard-boiled eggs, chopped
175g/6oz full-fat soft cheese
2tbls brandy
2tbls finely chopped parsley
salt and pepper
pinch each of cayenne pepper, nutmeg and allspice
Melba toast, to serve

- ***Serves 6***

1 Melt half the butter in a frying pan. Add the onion and fry for about 6 minutes, stirring occasionally, until transparent. Remove with a slotted spoon. Melt the remaining butter in the frying pan, add the chicken livers and cook, stirring constantly, over medium heat for 5-8 minutes. They should be tender with the insides still pink. Drain on absorbent paper.

2 Purée the onion and chicken livers with the hard-boiled eggs in a blender or food processor, a little at a time, at high speed until smooth.

3 Work the full-fat soft cheese and brandy together with a wooden spoon until light and fluffy.

4 Stir the chicken liver mixture into the cheese mixture. Add the parsley, salt and pepper to taste, cayenne pepper, nutmeg and allspice and stir well. Chill, covered, until ready to serve.

5 Serve the pâté accompanied by plenty of Melba toast.

GOING SOLO

ANJOU ROUGE

A glass of red wine as an aperitif? Why not? Usually thought of as best drunk with food, light and fruity reds shine out on their own

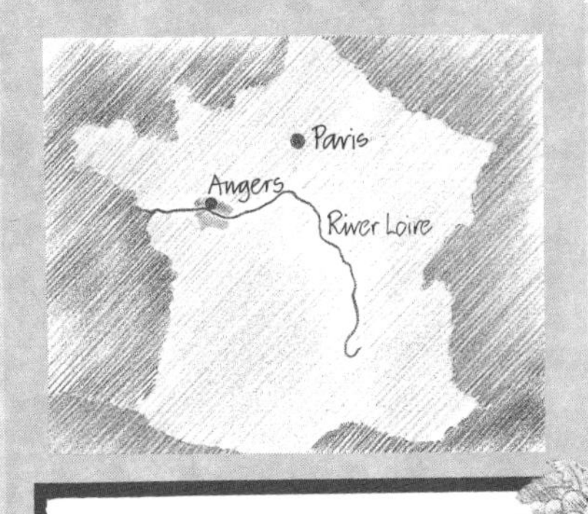

BEST BUYS

Anjou Rouge
Taste guide: *medium bodied*

Gamay
Taste guide: *light*

Côtes du Ventoux
Taste guide: *light*

Bardolino Classico
Taste guide: *light*

Syrah, Vin de Pays des Collines Rhodaniennes
Taste guide: *medium bodied*

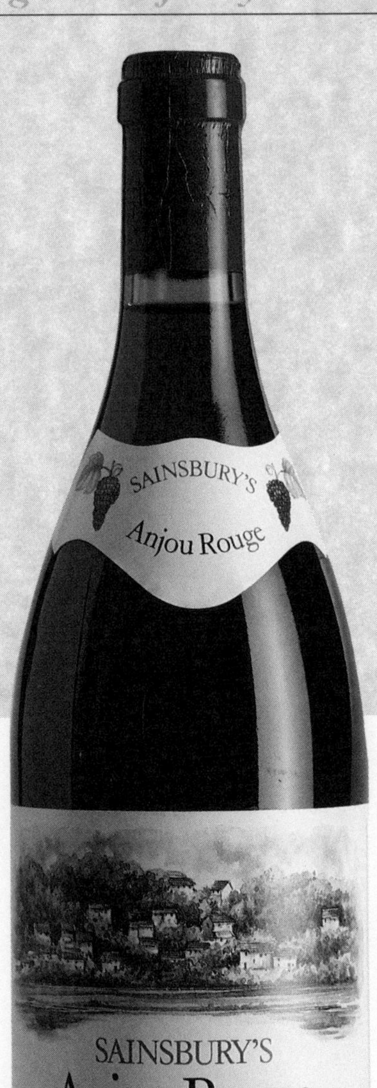

USUALLY, WHEN THERE'S the prospect of a nice, relaxing glass of wine in view it is assumed that it will be white, or possibly rosé. But why not red? Certainly many red wines are too full-bodied or have too much tannin (as in tea – it's the substance that dries your mouth out) to be at their best without food, although there are no hard and fast rules: you can and should drink anything you want on any occasion you choose if you personally enjoy it. But there are several styles of reds which are particularly well starred for drinking on their own, usually those which are lighter and crisper than the norm. A good test, though not an infallible one, is this: if the wine tastes at least as good, if not better, slightly chilled than at normal temperature, then it is likely to be good served without food.

The most obvious contender is Beaujolais. It's strange that the first cool red wine most people have is Beaujolais Nouveau in chilly November. But almost all Beaujolais is fresh, zippy and fruity and a joy to quaff without need for sustenance any time of the year.

THE CHILL FACTOR

Clearly temperature plays a significant part in the enjoyment of these wines, as it does any wine. It's often said that red wines should be served at room temperature, which is true for most of them but certainly isn't always the case. Light reds, those with little tannin and an abundance of fruit, are often better a little cooler – and the hotter the weather the cooler they can go, especially since they will warm up in the glass.

But for fuller-bodied reds, chilling is less successful; they can taste hard and metallic. So when is a room at room temperature? Basically it means pre-central heating, spring/autumn room temperatures, that is, neither warm nor cold to the lips. Just leaving the bottle in a niche away from radiators, fires, cookers and cold draughts should get it effortlessly to the right temperature.

Short cuts to warming a bottle of icy red don't generally work; the wine often takes on a nasty baked taste. The safest is to pour it straight into gently warmed glasses and leave in a warm – not hot – place or cup your hands round them.

At least too-cold red wine can be warmed. Once a bottle has become too hot, much of the lovely bouquet evaporates and the wine is spoilt for ever. Beware!

Soft ripe Gamay

Gamay is the grape variety which makes Beaujolais. One source is the Ardèche, which lies well south of the Beaujolais territory. The warmer climate of the Ardèche has resulted in a softer, riper wine, yet **Gamay** from the Ardèche retains all the liveliness and simple, fun gluggability that is the hallmark of this rightly popular wine.

Wines for chilly days

Nearby is Mount Ventoux and from the vineyards around it comes **Côtes du Ventoux,** with a bouquet like summer pudding and a gentle, comforting taste with pulpy fruitiness. Or, for the bolder, there's spicy **Syrah, Vin de Pays des Collines Rhodaniennes,** which tastes of blackcurrant-flavoured fruit gums. Syrah is the name of the grape which is found in most Rhône wines and makes long-lived, powerful wines like Hermitage. Designed to be drunk with rich meat dishes and cheeses, this wine is also surprisingly successful by itself even though it is quite full-bodied. Its low tannin and voluptuous fruit, however, compensate and it would be an ideal reviver on chilly or dull days. Its taste wouldn't be squashed either by flavoursome cheesy nibbles such as Cheese straws.

Fresh, fruity wines

Right at the other side of France, in the north west, lies the Loire valley and, around Angers, the district of Anjou. Better known for its rosé wines, Anjou also produces some very attractive reds which are well worth trying.

The grape in question is Cabernet Franc which is one of those responsible for the character of the revered wines of Bordeaux. But in the cooler climate and gentle atmosphere of the Loire it shines on its own and brings the aromas of summer – woodland raspberries and blackcurrants and a strong grassiness – to the wine.

Anjou Rouge is at its best served chilled: from the fridge it is surprisingly refreshing and alert, with bags of fresh, coolly grown raspberry-flavoured fruit, more grassiness, and plenty of clean up-front delicious vinous-fruity flavour.

Also better served chilled is **Bardolino Classico**. Bardolino is one of Italy's better known red wines produced in large quantities in an area near Verona, on the shores of Lake Garda. 'Classico' means the wine comes from the heartland of the region; traditionally the better part producing more 'classic' wines. Made from grapes rarely seen outside the area - Corvina, Rondinella and Molinara - Bardolino is often best when young and fresh. Sadly, however, too many producers keep it over-long before releasing it and too many wine shops age the wine even further, spoiling its vigorous but light-bodied attack. From a large chain with a quick turnover these problems can be avoided. Young Bardolino Classico has a cheery, youthful ruby-purple colour. It smells just like rich maraschino cherries and has all the power, fruit and flavour that only a red wine can give, together with all the invigorating chirpiness that is usually reserved for whites. That's what makes it so good to serve on its own, or with a crisp, cheesy nibble such as cheese straws (see below).

PERFECT PARTNERS

Cheese straws

- ***Preparation: 25 minutes, plus 30 minutes chilling***
- ***Cooking: 10 minutes***

50g/2oz butter, plus extra for greasing
100g/4oz flour
salt
pinch of cayenne pepper
65g/2½oz mature Cheddar, grated
1 large egg yolk
beaten egg, to glaze
2tsp poppy seeds
2tsp sesame seeds

- ***Makes about 100***

1 Heat the oven to 200C/400F/gas 6. Grease two baking sheets. Sift the flour, a good shake of salt and the cayenne pepper into a large mixing bowl. Cut the butter into the flour and rub in until it resembles breadcrumbs. Stir in the cheese. Beat the egg yolk with 1tbls water and add to the rubbed-in mixture. Blend the mixture together. Add a further 2tsp water, if needed, to form a ball.

2 Turn onto a lightly floured board and knead quickly until smooth. Wrap the pastry in stretch wrap and chill for 30 minutes. On a floured board, roll the pastry thinly to a 30cm × 25cm/12in × 10in rectangle. Trim the edges and cut the rectangle in half. Brush with the beaten egg and sprinkle one half with the poppy seeds and the other with the sesame seeds.

3 Cut the pastry into thin strips, about 5mm/¼in wide, then cut each strip into 6.5cm/2½in lengths. Twist the two ends of each straw in opposite directions and place on the greased baking sheets, spacing them 5mm/¼in apart.

4 Bake for 10-12 minutes or until golden brown. Cool on a wire rack before serving or storing for up to 2 days in an airtight container. If you want to serve them hot, refresh for 1-2 minutes in an oven heated to 200C/400F/gas 6.

WELL-DRESSED SALADS

TOLLEY'S PEDARE 1988

The creamy texture of mayonnaise-dressed salads demands the firmness of a lively white wine.

THERE ARE INNUMERABLE occasions when salads are served: as a meal in themselves, as a simple side salad to accompany cold meats, as part of a buffet. Often the salads will be dressed with mayonnaise, or at least a dressing based on mayonnaise. And on most of these occasions, wine will be served. It is not too hard to find a wine, preferably white, that will go well, although some careful selection is needed.

Creamy on the tongue

Despite the many different varieties of salads that can be made, using all sorts of ingredients, there are basically two different types – at least as far as choosing wines to match is concerned. The first is where the mayonnaise acts as a sauce to amalgamate and tone down the flavours of assertively flavoured and often quite acidic vegetables and fruits, such as peppers, onions, sweetcorn, apple, celery and so on. In this case, the mayonnaise will probably be quite gently flavoured and based on a light vegetable oil. The other type of salad is where the salad ingredient itself is comparatively lightly flavoured; pasta or potato, for example, and the mayonnaise is the thing. If making this sort of salad, chances are you will go for a pungent, fully-flavoured mayonnaise based on good olive oil (preferably extra virgin). Depending on which type of salad you are serving, the wine you choose will tend to vary. For a range of salads of both types have wine for both.

A friend from Down Under

One wine that does match both is **Tolley's Pedare 1988 South Australia Chenin Blanc Colombard**. Chenin Blanc and Colombard are the grape varieties from which the wine is made, both of them native to France, but both of them well suited to South Australia where the warm climate ripens them well and gives them extra weight. Tolley's is zingily fresh, with an enlivening bouquet of lemons, limes and persimmons. It tastes just as mouthwateringly crisp and fruity and has a good lemon and greengage tang and a slight saltiness.

BEST BUYS

Tolley's Pedare 1988 South Australia Chenin Blanc Colombard
Taste guide: dry

Cooks Marlborough Sémillon 1989
Taste guide: dry

Château Roquetaillade La Grange 1987
Taste guide: dry

Les Morlières Muscat 1988 Vin de Pays de l'Hérault
Taste guide: dry

Alsace Pinot Blanc Cave Vinicole Eguisheim
Taste guide: dry

Soave Classico i Ciliegi 1988 Lamberti
Taste guide: dry

With the vegetable combined with fruits in mayonnaise salad it tastes a little softer and rounder and with the mayonnaise-led salads it provides a delicious contrast of flavours.

Another good match is **Cooks Marlborough Sémillon 1987**. It comes from New Zealand, a country whose wines are becoming more and more popular – and rightfully so. The climate, particularly around Marlborough in the north of South Island, is ideal for many white grape varieties, including the Sémillon. The ripe, almost sweet character of the fruit softens the attack of the mayonnaise-based vegetable and fruits salads, while its own fruity acidity is a lively complement to potato and pasta-type salads.

Also from the Sémillon grape variety, but this time from its original homeland of Graves, in the southern part of the Bordeaux region, is **Château Roquetaillade La Grange 1987**. It has a most attractive, delicate, creamy nutty bouquet with a rounded, harmonious mellowness that comes from the oak casks in which it has been matured. It goes better with the less sharp vegetable-in-mayonnaise type salads than with the more strongly assertive flavours, but it really comes into its own when a mayonnaise is the most important salad ingredient.

A rare dry Muscat

If you are preparing pasta, potato or rice-based salads, then **Les Morlières Muscat 1988 Vin de Pays de l'Hérault** is well worth considering. Although the Muscat grape is probably better known for producing sweet wines, it can make some remarkable dry wines too, although even these give the sensation that they want to be sweet. A good, mayonnaise adds an extra dimension to its flavour.

Alsace Pinot Blanc from the Cave Vinicole Eguisheim is another wine well worth trying with mayonnaise-dressed salads. Alsace makes wines, usually from single grape varieties such as Pinot Blanc, which are certainly among the best in all France, possibly in the whole world, for partnering with a wide variety of foods. Its proximity to Germany results in wines with plentiful aroma, but the French influence brings plenty of weight and firmness. A gentle, creamy-salty bouquet and a creamy, nutty, crisp flavour is a great match with a potato salad and not bad at all with a vegetable-based one.

And Soave's reputation as the buffet wine *par excellence* was not gained for nothing. Just try **Soave Classico i Ciliegi 1988 Lamberti** with as wide a range of mayonnaise-dressed salads as you like. Its gently nutty flavour acts as a gentle base to all of them and it slips down well. If you make mayonnaise, you can balance the acidity and, to a certain extent, adjust the flavour to suit the wine. The easiest way to make it is to use a food processor. Break 2 eggs into the bowl of the processor, add ½tsp French mustard and 2tsp lemon juice and switch on. Pour in 300ml/½pt oil in a steady stream until the mixture is thick. Season to taste with salt and freshly ground black pepper and add more lemon juice or mustard if necessary.

PERFECT PARTNERS

Pimento potato salad

- ***Preparation: 30 minutes***
- ***Cooking: 15 minutes***

1kg/2¼lbs new or waxy potatoes
salt
300ml/½pt vinaigrette dressing (see page 28)
1 garlic clove
300ml/½pt mayonnaise (see above)
2 canned pimentoes, finely chopped, liquid reserved
1tsp paprika
To serve
1 lettuce, washed and dried
4 tomatoes, cut into water lilies
1 hard-boiled egg yolk, sieved
1 bunch watercress
paprika

- ***Serves 6***

1 Several hours before the salad is needed, scrub the potatoes and cook in boiling salted water until cooked through but still firm.

2 Drain the potatoes reserving 2 of them. Whilst still hot, remove the skins from the remainder and dice them. Put into a large bowl and toss in the vinaigrette. Refrigerate for at least 2 hours with the tomato lilies.

3 Crush the garlic clove with a little salt and mix into the mayonnaise. Reserve 2tbls of this to use later. Mix the finely chopped pimentoes, their liquid and the paprika into the rest of the mayonnaise.

4 One hour before serving, toss the cold potatoes into the flavoured mayonnaise and refrigerate till needed.

5 Just before serving, line a dish with the lettuce leaves and pile the potato salad on top. Fit a piping bag with a star nozzle. Sieve the 2 potatoes into the 2tbls mayonnaise and mix well. Pipe the mixture into the tomato halves. Sprinkle with the sieved egg yolk around the salad.

SHARP DRESSERS

SAUVIGNON HAUT POITOU

Standing up to a vinaigrette dressing is a serious challenge for even the best wine

WHEN THINKING ABOUT wines to go with a meal, it is all too easy to forget that important item – salad. It may be an accompaniment to the main course, an integral part of one or more dishes or it may be the basis of the entire meal. A salad may be as simple as plain dressed lettuce or it may involve a large number of ingredients varying widely in flavour and texture. Is it reasonable to expect to find wines which will go with all sorts of salads? Of course not, but it is reasonable to look for wines that will go with the majority of them, especially since there is one ingredient in most of them that is often the arbiter of wine-matchability: a well-balanced vinaigrette dressing.

Oil and vinegar

Vinaigrette is the best way yet devised for blending the flavours of various salad ingredients. The oil is the major unifier, the vinegar highlights and seasons the flavours.

Unfortunately, vinegar is not a great friend of wine – calling a wine 'vinegary' is a stern criticism. Nevertheless, if the vinegar is in the right proportion (about one part to three of olive oil) and if the salad isn't drenched in dressing, it is possible to find a number of wines not adversely affected by it. Remember that if you plan to use a very special young, extra virgin oil, which can be very peppery, or a nut oil, or one of the fruit or herb vinegars, you should try out your chosen wine with the salad before you offer it to guests – make sure of the flavours.

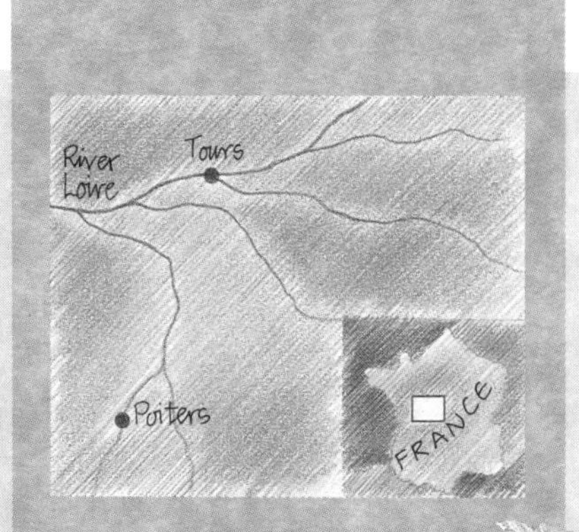

BEST BUYS

Sauvignon Haut Poitou 1988
Taste guide: very dry

Vin de Pays de Loire Atlantique 1988
Taste guide: very dry

Château Bonnet 1988 Entre-Deux-Mers
Taste guide: dry

Gewürztraminer 1987 Scherer
Taste guide: dry

Le Viole Bardolino Classico 1987 Lamberti
Taste guide: light-bodied

White and crisp and even

For a light, easily quaffable wine which will suit a wide variety of salad ingredients, try **Vin de Pays de Loire Atlantique 1988**. Remarkably good value, it is made from the Muscadet grape variety though it comes from a wider area than Muscadet, from the département around the mouth of the Loire in western France. It is as crisp as Muscadet but less penetrating in flavour; it stands up well to vinaigrette.

The Sauvignon grape variety makes delicious fresh wines which surprisingly do not go well with as many foods as you might think. But they are just right with all sorts of salads because their crisp acidity stands up well to the vinegar part of the dressing. Try **Haut Poitou 1988 Sauvignon**: it comes from western France, just south of the Loire and north of Poitiers. Its enlivening grassy, gooseberry-like bouquet and fine, dry, crisp taste are an ideal match for buffet meals, main course salads, and side salads (especially with a fish course). In fact, it will sit happily with any vinaigrette salad.

Extra from the oak

Sauvignon takes on quite a different and sometimes quite superb style if left to age in oak barrels for a time. A great example of this is **Château Bonnet 1988 Entre-Deux-Mers**, made by André Lurton. Vanilla and nuts flavours from the oak mingle with the green fruit of the grape to make a refined complex wine whose Sauvignon base will handle vinaigrette dressings with ease.

Something from Alsace

If salad is the basis of your meal and if you have some pretty assertively flavoured ingredients in it, apart from the dressing, you are going to need a pretty assertively flavoured wine to cope with it all. This could be the occasion for a **Gewürztraminer 1987 Scherer.** The spicy Gewürztraminer grape, grown in Alsace, can produce big, powerful wines but can also produce rich, not overly weighty, perfumed wines like this one – a wine with a restrained hint of lychees and violets whose spiciness is less dominant than usual. Its intensity of flavour is just right to ensure that the vinegar doesn't take over.

Your salad may include meat or be a side dish to meat, so you may prefer a red wine. If so, then **Le Viole Bardolino Classico 1987 Lamberti** will fit the bill. From the eastern shores of Lake Garda in Italy, it is a lightish red colour with a fragrant meat-stock bouquet and has a lightweight cherry-fruit taste with low tannin and a clean, dry finish.

The fruitiness of Alsace wines would go well with dressings made with other vinegars, too. You may like to experiment with either red wine vinegar or sherry vinegar, or, for a change, make your dressing with a fruit-flavoured vinegar such as raspberry, or the rich brown, and wonderfully sweet-and-sour-flavoured, balsamic vinegar.

PERFECT PARTNERS

Summer salad

- ***Preparation: 20 minutes plus chilling***

1 large lettuce
½ curly endive
For the vinaigrette:
½ tsp Dijon mustard
2 tbls wine vinegar
salt and freshly ground black pepper
6 tbls olive oil
For the garnish:
1 large tomato, cut into 8 wedges
12 black olives, stoned and soaked in olive oil
2–4 tbls chopped mixed fresh herbs – parsley, chives and tarragon

- ***Serves 4*** 

1 Wash the lettuce and curly endive, cutting off and discarding all coarse or damaged leaves. Drain well, then pat the leaves dry in a clean cloth. Lay out the leaves on a fresh cloth, roll up the cloth and chill in the salad compartment of the refrigerator until you are ready to assemble the salad.

2 To make the vinaigrette, put the mustard in the bottom of a small jug or cup and add the vinegar. Add salt, black pepper to taste and olive oil. Beat vigorously with a fork or whisk to make an emulsion. Alternatively make the dressing in a screw-top jar.

3 When ready to serve tear the salad greens into large pieces and toss lightly in a salad bowl.

4 Pour the vinaigrette over the salad greens and toss until each leaf is glistening. Arrange the tomato wedges and black olives decoratively on top of the salad. Sprinkle the salad greens with chopped fresh herbs and serve immediately with a main course.

FRESH STARTS

MUSCAT BEAUMES DE VENISE

Even the best wines face stiff competition with citrus fruit starters

IT MAY BE that sometimes we expect too much of wine for we tend to think that there just must be at least one wine somewhere that will be the perfect partner for any dish you care to conjure up. For the most part this is true – but sometimes we eat foods that really put this theory to the test. One of the hardest to find a good match for is citrus fruit.

Citrus fruit starters have many benefits. They are light, refreshing, appetising, enhance rather than sate the appetite and have obvious benefits for slimmers. Finding the right wine, though, can be really quite a trial. What is needed is something white that is not too acidic, and preferably sweet, as the fruit's acidity makes everything consumed with it appear dryer. Yet, at the beginning of a meal, the sort of wine you are looking for is probably crisp, light and dry – exactly the wrong sort of thing. One way to help is to go easy on the grapefruit, which can leave a bitter sensation in the back of the throat. If possible, choose ruby grapefruit, which are sweeter and less bitingly acidic-bitter.

Muscatel magic

The best style to complement citrus fruit is epitomised by **Muscat de Beaumes de Venise 1987.** It is strong (15% alcohol), fully flavoured and pretty sweet. In fact, it is the sort of wine which you would more usually drink with dessert or after a meal. But it certainly partners orange, grapefruit, mandarin and so on. One reason is the Muscat grape variety from which the wine is made and which often has its own flavour of orange peel. This wine has a floral, citrus oil and candied fruits bouquet. It is very fresh, but not all that high in acidity, with lots of ripe grapey-orangey fruit. It is powerful enough not to be overwhelmed by the fruits' tastes and yet allows the clear taste of the fruit to show through. If you normally add sugar to your citrus fruit, you can add a little less than normal with this wine, or with any sweet wine, as the after-taste of the wine in your mouth acts as its own natural sweetening agent. But do make sure you have a good few sips of water before going on to a dry white or red later in the meal – or else it will taste unpleasant and very strange indeed!

BEST BUYS

Muscat Beaumes de Venise 1988 (two glasses)
Taste guide: very sweet

Arione Asti Spumante
Taste guide: sweet (one glass)

Château Moncontour 1986 Vouvray (two glasses)
Taste guide: medium dry

Alsace Gewürztraminer 1988 (two glasses)
Cave Cooperative d'Ingersheim
Taste guide: dry

Domaine de Planterieu 1988 (one glass)
Vin de Pays des Côtes de Gascogne
Taste guide: dry

Frothy sparkle

While sticking with the Muscat grape and still going for a sweet wine, you could choose something from completely the other end of the weight spectrum – **Arione Asti Spumante**, coming from south of Turin in north-west Italy. It shows the grape variety at its delicate best. This sparkling wine has only half the alcohol (7.5%) of the Beaumes de Venise, is full of gentle frothiness, with grassy, grapey aromas and is a mouthful of grapey delight. It sets off the tastes of mandarin, clementine or satsuma perfectly, although it is too light for the most sharply and penetratingly flavoured grapefruit.

A wine that breaks another of the rules for good citrus fruit combination, but gets away with it, is **Château Moncontour 1986 Vouvray.** From the Loire valley of northern France, it is made from the Chenin Blanc grape. Characteristically for that variety, it has a gentle bouquet of honey and flowers. It is only medium-dry, but also usually with good acidity. On this wine, the acidity isn't too high but it does have a slight grapefruity aftertaste, which helps explain why it manages to cope with the citrus fruit. In fact, it tastes softer, less crisp and drier, noticeably different, but still very enjoyable.

If you feel that you'd prefer to avoid a sweet wine early on in a meal, the solution could be **Alsace Gewürztraminer 1988** from the Cave Cooperative d'Ingersheim. It has the classic spicy lychees bouquet of the Gewürztraminer grape and a mouth-filling, oily, rich, spicy taste with comfortingly low acidity. It's dry but not too dry. Gewürztraminer can make blockbuster wines but this example, although fully-flavoured and full-bodied, isn't too weighty. Its body helps it to balance all types of citrus fruit and it is particularly good with oranges, bringing out their flavour.

Just a touch of dryness

Is it possible to find a crisp, dry wine that would be fine as an enjoyable glass before you sit down to eat which will still be enjoyable with grapefruit? Just about. **Domaine de Planterieu 1988 Vin de Pays des Côtes de Gascogne** with its fresh, grassy, apple, nutty flavours is so refreshing and vibrantly fruity that it almost acts as another citrus fruit, so combines quite happily with those you are eating. It *is* slightly overwhelmed by a mouthful of grapefruit or orange. And if you sprinkle a few nibbed nuts on your starter, the wine's nuttiness will be an added match.

Another way of softening the sharp flavour of citrus fruits is to combine them with other ingredients. In the fennel, grape and orange salad below, the sharp, tangyness of the orange is offset by the quite strong, nippy flavour of the watercress, and balanced with the cool, crisp taste and texture of both cucumber and fennel. Grapes add their own, fruity sweetness to make a delicious citrus fruit salad to serve with either a medium-dry or dry wine.

PERFECT PARTNERS

Fennel, grape and orange salad

A crisp combination of vegetables and fruit, this salad is particularly good with braised pheasant or duck.

- ***Preparation: 20 minutes, plus 1 hour standing***

1 bunch watercress
2 fennel bulbs, thinly sliced
2 large oranges, peeled and segmented
10cm/4in length of cucumber, peeled and thinly diced
100g/4oz black grapes, seeded
4tbls sunflower oil
1tbls lemon juice
1tbls red wine vinegar
pinch of caster sugar
salt and freshly ground black pepper

- ***Serves 4*** (!) (££)

1 Remove any yellow leaves from the watercress, wash thoroughly and divide into sprigs.

2 Mix together the sliced fennel, orange segments, diced cucumber and grapes. Reserve a few watercress sprigs for garnish and mix in the remainder.

3 Place the sunflower oil, lemon juice and vinegar, sugar, salt and pepper in a screw top jar, shake well to mix. Pour the dressing over the salad and cover. Stand at room temperature for 1 hour.

4 Turn the salad into a serving dish, garnish with the reserved sprigs of watercress before serving.

ASPARAGUS MATTERS

ORLANDO JACOB'S CREEK 1988

The distinctive flavour of asparagus is a real problem-setter for all wines

THERE ARE SOME foods that are notoriously difficult – some would say impossible – to partner well with wine. Chocolate is one, artichokes another, and so is asparagus. Yet asparagus is such a fine food and makes such a perfect starter that it seems hard to believe there isn't at least one wine that will show it off to perfection. In fact, there is no need to abstain from wine completely when eating asparagus – there are wines to suit it. True, many will taste thin and tinny but there are some that make a perfectly respectable match.

Asparagus changes the taste of *all* wines, though, so if you like a certain wine on its own, even if it tastes a little asparagus-like (some do!), don't expect it to taste the same with asparagus – it won't. And don't serve it to guests without trying out the partnership first. But don't worry about the wine changing the taste of your asparagus – it is far too incisively flavoured to be affected by any mere liquid! Don't waste good money on an expensive wine unless you are quite sure it will go. Asparagus tends to denude wines; posh bottles may taste good enough with it – but not as good as without it – and you are losing the taste of the fine fruit and subtleties that you have paid for.

Play safe with white

One wine that fulfills all the criteria for drinking with asparagus is **Mâcon Villages 1987 Maurice Chenu**. It is white (reds are disastrous with asparagus!), comes from Burgundy and is a well-known and well-regarded name, although it is not one of the really expensive top-notch Burgundies. This particular wine has more character than most Mâcons. A bright, deep straw colour, it has a lovely silky feel to it with a rich, almost voluptuous, buttery, appley, earthy bouquet and a soft, rounded, refined character. It loses a bit of its fruit with asparagus – but not much – and in fact it stands up well to the partnership, becoming more refreshing.

BEST BUYS

Orlando Jacob's Creek 1988 South Australia Semillon Chardonnay
Taste guide: dry

Mâcon Villages 1987 Maurice Chenu
Taste guide: dry

Lindemans 1989 South Eastern Australia Chardonnay Bin 65
Taste guide: dry

Ritzenthaler 1988 Alsace Tokay-Pinot Gris
Taste guide: dry

Caliterra 1989 Curicó Chardonnay
Taste guide: dry

Australian treats

Mâcon Villages is made from the Chardonnay grape, as is **Lindemans 1989 South Eastern Australia Chardonnay Bin 65**, from half a world away. The Bin Number is quite important. Lindemans is a large company producing high quality wine in Australia. They make numerous different styles, types and qualities of wine, and to distinguish one from another they give each a Bin Number. So, if you see another Chardonnay from Lindemans with a different Bin Number it may be better, worse or as good as Bin 65, but it will certainly be different. Bin 65 is a very powerful wine. It has all the heady richness of wine made from really ripe grapes grown in Australia's warm countryside, and it has been given extra weight and oaky, vanilla-like characteristics from a short stay in oak barrels. The weight and oakiness is more apparent on the bouquet than in the mouth; once sipped, the full-flavoured fleshy, butteriness of the Chardonnay grape is dominant, with the oakiness a secondary characteristic. Good as this wine is, it can seem almost too weighty for drinking on many occasions. That is why it works so well with asparagus, for the asparagus makes it seem much less powerful, although lets most of its fruit come through.

Even better, though, is **Orlando Jacobs Creek 1988 South Australia Sémillon Chardonnay**. It still comes from Australia and there is still the Chardonnay grape used, but this time the Chardonnay is blended with some Sémillon, which lightens the wine and gives it more elegance and balance. Sémillon is a grape (French in origin) which has found its true niche in Australia, producing wines well worth trying, but not necessarily ones that go well with asparagus. The combination between asparagus-suiting Chardonnay and taste-enticing Sémillon is, however, an unbeatable one and makes the wine a winner. With asparagus, it loses a bit of its weight but gains in return a crisp steeliness that helps clean the mouth after the clinging flavours.

A partner from Alsace

Since wines from Alsace, on the east French border with Germany, are in general such good food partners, it seems reasonable to expect that there will be at least one wine from Alsace that will be able to cope with asparagus. There is: **Ritzenthaler 1988 Alsace Tokay-Pinot Gris**. Tokay and Pinot Gris are simply different names for the same grape variety, but because there was a risk that Tokay could be confused with a totally different wine from Hungary, from a different grape, called Tokaji (pronounced Tokay) it was decreed that the Alsatians could not use its name without its synonym too. Because the flavour is so long lasting, it doesn't let the asparagus get the upper hand.

As one more example of the ability of the Chardonnay grape variety to cope with asparagus, try **Caliterra 1989 Curicò Chardonnay** from Chile. It is a fatter, broader, more barley sugar-like style of Chardonnay with an attractive tang.

An even more complex partnership problem occurs when you combine asparagus with another strong flavour, such as a tangy dressing or a flavourful sauce, as is the case with the asparagus salad below. So before you make your choice, refer first to pages 25-28 to find out more about wines to go with sharp - and not-so-sharp - dressings.

PERFECT PARTNERS

Asparagus salad

- ***Preparation: 20 minutes***
- ***Cooking: 8-20 minutes depending on thickness***

1kg/2¼lb fresh asparagus
salt
12 large sprigs of fennel
For the soured cream dressing:
150ml/5fl oz soured cream
1tbls lemon juice
salt
freshly ground black pepper
1tbls finely chopped parsley
1tbls finely snipped chives
1tbls finely chopped tarragon
dash of hot pepper sauce or pinch of paprika
1tsp snipped chives, to garnish
1tsp chopped parsley

- ***Serves 4***

1 To prepare the asparagus, trim off any woody part from the stalks. Cook in a pan of boiling salted water for 8-20 minutes, depending on thickness, until tender.

2 Drain the asparagus and refresh under cold running water. Drain again and lay on a folded clean tea-towel to absorb any remaining water. Leave to cool.

3 In a bowl, combine the soured cream and lemon juice. Season with salt and freshly ground black pepper to taste. Add the fresh herbs and a dash of hot pepper sauce or a pinch of paprika, and stir until the dressing is smooth and well-blended.

4 To serve, lay 3 spr gs of fennel on each plate. Divide the asparagus into 4 portions, and arrange on the fennel sprigs on each plate. Spoon 2tbls soured cream dressing on the side of each plate and serve the remaining dressing separately, garnished with a little snipped chives and chopped parsley. This salad looks most attractive when served on asparagus plates.

HANDY HALVES

CHARDONNAY

Half bottles are hard to find, yet they are ideal for so many occasions

IT'S IRRITATING NOT to find more wines in half bottles. The size is ideal for so many purposes, and you would have thought that those in the forefront of the movement to persuade us to 'drink less but drink better' would have been campaigning for shops to stock far more wines in halves. But, at least for the foreseeable future, we have to be content with the few that are available.

A bite of citrus

Take, for example, **Chablis 1987,** produced by Remy Le Fort. Chablis is a fish wine par excellence, especially with white fish like sole or plaice, served with a light, butter sauce. That's because Chablis comes from the buttery Chardonnay grapes of the Burgundy region. But, unlike many white Burgundies, Chablis, coming from the north of the region, is less fat and crisper, with a good burst of acidity, which has much the same effect as squeezing lemon juice over fish. Sadly Chablis as a name is all too popular, and far too many wines, despite being fairly costly, don't live up to its reputation. This one does. It is pale in colour, with a slight green tinge that is characteristic.

Aussie revolution

A really pleasant rarity is to see an Australian wine in half bottles. The appearance of so many Aussie wines on the shelves has been the wine revolution of the late '80s and one of the most beneficial revolutions any wine drinker could hope for. Most are ripe, full-bodied, fruit-packed wines with strong flavours of the grape varieties from which they are made and are often little short of sensational. Their only possible slight disadvantage is that they are at times just a little too powerful for you to want to drink more than a glass or two. That is when the half bottle comes into its own. **Orlando RF Southeastern Australian Chardonnay 1987** is the one to track down. Orlando's is one of the most successful Australian Chardonnays around and is a bright deep gold in colour. As it is from the Chardonnay grape it also has a buttery taste and smell, but different ones from Chablis. Try it when you make a chicken salad.

Small sweet sips

Half bottles come into their own with sweeter wines. The very sweetness means that their flavours seem

BEST BUYS

Orlando Australian Chardonnay
***Taste guide:* dry**

Chablis 1987
***Taste guide:* very dry**

Friedelsheimer Schlossgarten 1984 Huxelrebe Auslese
***Taste guide:* sweet**

Domaine de Coyeux 1986
***Taste guide:* very sweet**

Château Bastor-Lamontagne 1986 Sauternes
***Taste guide:* very sweet**

to coat the mouth and stay there, so you need take only small sips to enjoy their taste. Not only that, the usual role of a sweeter wine is either as an aperitif or at the end of the meal with dessert. In neither case are you likely to want more than a glass, so it can seem wasteful to buy a whole bottle.

Despite the ease of finding half bottles of German wine the finer, higher quality wines – which are also the sweeter ones – are comparatively rare. Yet it is well worth while trying one. It is like a glass full of all the best things in a normal German wine but far more concentrated. A half is the ideal way to make the acquaintance: **Friedelsheimer Schlossgarten 1984 Huxelrebe Auslese** is a model example. It comes from the village of Friedelsheim in Germany's Rheinpfalz district, where the wines are often comparatively big and rich. It is also from the Huxelrebe grape variety, which imparts a gentle spiciness. This wine is quite simply all that it should be and has an extra raisiny edge.

Spirited Muscat

A superb dessert wine (though the French love it as an aperitif too) is Muscat de Beaumes de Venise. To see it at its best try **Domaine de Coyeux 1986.** It is made from the highly adaptable Muscat grape, which can make sweet or dry, light or heavy, sparkling or still, or any other type of wine and usually leaves a real grapiness so that drinking the wine can have almost the same effect as eating fresh Muscat grapes. Beaumes de Venise is a fortified Muscat wine, a little grape spirit being added to it when it is almost fermented. This strengthens it and stops fermentation, so the wine remains sweet. It has an alcohol content of 15% and is best drunk in small quantities. Sauternes is one of the most prized dessert wines around. It is not easy nor cheap to produce because it can be made only when there are certain weather conditions in autumn, with sunny days but misty mornings and evenings, encouraging the growth of a special fungus which pierces the grape and sucks out the water, concentrating its other elements and slightly changing its characteristics - for the better. With a fruit-based dessert, such as lightly poached pears, peaches or nectarines; flavourful fruit pies (either hot or cold) made with apples, plums or cherries; a simple, chilled salad made with fruits such as melon, raspberries, strawberries and peaches; or simply one or more of these summer fruits served raw as a refreshing finish to a meal, Sauternes would be excellent. **Château Bastor-Lamontagne 1986 Sauternes** is quite strong in alcohol too, but isn't 'sticky' sweet and leaves a clean taste in the mouth.

PERFECT PARTNERS

Chicken and nasturtium salad

- ***Preparation: 45 minutes***
- ***Cooking: 1 hour, plus cooling time***

20 nasturtium leaves
2 bouquets garnis
½ lemon, thinly sliced
1.6kg/3½lb roasting chicken
1½tsp ground ginger
1½tsp turmeric
1 small onion, roughly chopped
1 small carrot, roughly chopped
1 celery stalk, roughly chopped
4 cloves
1tsp black peppercorns
150ml/5fl oz soured cream
2tsp chopped capers
1 garlic clove, crushed with a pinch of salt
½ medium-sized cucumber
1 medium-sized lettuce, shredded
4 nasturtium flowers, to garnish

- ***Serves 4*** (ff) (££)

1 Put 2 nasturtium leaves, a bouquet garni and the sliced lemon inside the chicken. Truss it and rub 1tsp each of ginger and turmeric into the skin.

2 Put the chicken into a large saucepan with the onion, carrot, celery, cloves and peppercorns. Pour in just enough water to cover the legs, but not the breast.

3 Set the pan over a medium heat and bring to the boil. Skim, cover and simmer gently for 50 minutes, or until the chicken is really tender.

4 Lift out the chicken and cool. Strain the stock. Bone the chicken and dice.

5 Pour the soured cream into a bowl and beat in ½tsp each ginger and turmeric. Finely chop the remaining nasturtium leaves and mix them in, along with the capers and garlic, fold in.

6 Cut the cucumber into quarters lengthways, then thinly slice the quarters. Arrange a bed of lettuce and cucumber on a serving dish, top with the chicken salad and garnish with flowers.

EVERYTHING'S COMING UP ROSES

GRENACHE

Rosé wines are often overlooked, considered the poor relation of white and red, but a rosé holds its own as the perfect drink with crab

THERE ARE GENERALLY two types of rosé wines. The first is rather like a white wine with a bit of red character. The second is more like a red wine with a bit of white character. The first is for when you feel like the crispness and lightness of a white, but with just a little more body; the second is for when you fancy the fruit of a red, but want something a little more delicate and refreshing.

Skin deep

Don't imagine, though, that rosé wines are made by mixing red and white – they never are. The colour in all wines depends on the skin of the grape. In most grapes, black and white, the juice is white. When grapes are fermented into wine, if they are just crushed, the juice mixes with the skins and extracts the colour. White wines are usually pressed quickly so hardly any skin contact occurs. Red wines keep juice and skins together for days, sometimes two weeks or more, to extract all the colour there is (and flavour elements too). Rosé wines are made from red grapes but the wine stays in contact with the skins for just a short time. The deeper the skins, the shorter the contact the juice needs for a given colour.

Pink and blue

At the Listel Domaine on the south coast of France, by the salt flats of the Rhône delta, the palest, most delicately coloured of rosés is made. It is so prettily light that the French don't even call it rosé, but Gris (grey). They go even further and call the wine Gris de Gris (grey wine from grey grapes). This is because when the grapes (Carignan, Cinsault and Grenache varieties) are harvested they have a bluish-greyish hue to them. The juice stays with the skins just 20 minutes: the time it takes to drain the large hopper into which they are loaded, and in that time it picks up just the right amount of colour.

Listel Gris de Gris is a pale coral colour. Sniff it and imagine the warmth of the south of France with a gentle sea breeze blowing. Its taste is light and refreshing, yet with your eyes closed it doesn't seem quite like a white: it has just a little more

FRANCE
River Rhône
Avignon
Nice
Agde
Cap d'Agde
Marseille

BEST BUYS

Cante Cigale 1987 Grenache
Taste guide: **dry, light-bodied**
Serve with: **salmon, crab**

Listel Gris de Gris
Taste guide: **dry, light-bodied**
Serve with: **shellfish**

Rosa dell'Erta 1987
Taste guide: **very dry, light-bodied**
Serve with: **shellfish, salmon trout, crumbly cheeses**

Chateau la Jaubertie 1987 Bergerac Rosé
Taste guide: **dry, light-bodied**
Serve with: **veal, richer fish dishes**

roundness and weight. It's a delightful aperitif, but with shellfish it really comes into its own.

Strawberry cup

Also from the south of France, this time the Cap d'Agde, is **Cante Cigale 1987 Grenache**. A pretty pale-rose pink, it smells, like so many rosés, of fresh strawberries. Once you take a sip the first sensation is of crispness, like a good apple, then the fruit and fullness take over and after swallowing it leaves a welcome sensation of warmth. A clue to the surprising amount of flavour in this wine, without its becoming heavy, is its grape variety, Grenache. Grenache usually makes full-blooded red wines, with plenty of power and 'oomph'. So even the short skin contact needed to get the pink tone will also extract plenty of flavour. Then there's the particular benefit of the volcanic soil of the Cap d'Agde to give it its individuality. Light enough for the most delicate fare, yet strong enough for the richness of salmon, it's a real bargain and well worth a try.

Rosy blush

Rosa dell'Erta 1987 is even more light and dainty. The label could not be more apt: a line drawing of a single rose. It comes from Tuscany, in Italy, from one of the best Chianti estates, Castello di San Polo in Rosso. It is left hardly any time at all on the skins of the red grapes (predominantly Sangiovese) usually used to make Chianti. The colour is quite a deep rose pink, despite its short contact time, which means the grapes must have been quite dark when picked. The style is very close to a white wine; only the touch of tannin and the dryness after swallowing give the game away. A mixture of roses and strawberries on the nose, crisp and dry in the mouth with gently strawberryish fruit, it is really quite a classy wine which would match - apart from shellfish and pink fish like salmon trout - light and crumbly cheeses as well.

Hidden charms

From the look of **Château la Jaubertie 1987 Bergerac Rosé** it could almost be a pale red wine. The colour is deep pink, like red cherries. The aroma is equally confusing: the ripeness and fruitiness of a red, the delicacy and strawberryishness of a rosé. On the taste, though, it has to be a rosé - and not too full a one at that - with plenty of gulpable fruitiness but also charm and freshness. Bergerac is in the south west of France, near Bordeaux, and makes wines that are somewhat similar in style to Bordeaux's red clarets, dry whites and sweet whites. Since rosé wines are still relatively unusual, those who do make them take a particular amount of care to make them really rather special.

PERFECT PARTNERS

Crab and cucumber salad

This dish is tastiest made with fresh crabmeat, but you can also use frozen or canned. Make sure that canned crabmeat is thoroughly drained

● ***Preparation: 40 minutes, plus 1 hour chilling***

6 spring onions
1 cucumber, cut into 1cm/½in chunks
225g/8oz white and brown crabmeat
For the dressing:
2tbls olive oil
2tbls light soy sauce
1tsp peeled and finely chopped fresh root ginger
1-2 tsp lemon juice
pepper

● ***Serves 4***

1 Finely slice the white part of the spring onions, reserving the green stalks, and mix with the cubes of cucumber. Arrange around the edge of four serving dishes. Pile the crabmeat in the centre of each dish, placing the brown crabmeat in the centre of the white. Garnish with shredded spring onion stalks and chill.

2 To make the dressing, combine all the ingredients in a small bowl, adding pepper to taste. Whisk with a fork until smooth. Transfer to a serving jug and hand round separately.

WINES WITH VICHYSSOISE

GUNTRUM 1987

BEST BUYS

Louis Guntrum 1987 Sylvaner Dry
***Taste guide:* dry**

Vin de Savoie Apremont Jean Cavallé
***Taste guide:* dry**

Alsace Pinot Blanc Cave Vinicole Eguisheim
***Taste guide:* dry**

Puerto Fino Superior Dry Sherry Burdon
***Taste guide:* very dry**

Creamy soups can blur the crispness of a wine, but excellent whites can enhance the richness of their flavour

IT IS NOT unusual to start a meal with a soup of some sort, yet it is quite unusual to bother about a wine to go with the soup. Perhaps, because soup is itself a liquid, there is a tendency not to drink wine with it and to start with wine for the main course. But there is no reason at all why soup, just as any other starter, should not be complemented by a glass of wine. And as there is hardly likely to be a need for much more than one glass, it can be the ideal way to finish off a bottle opened for an aperitif. All the wines recommended below are good aperitifs.

Vichyssoise is not the easiest of soups to partner with wine, though. Although it may be quite delicately flavoured, the taste of leek can be incisive and react badly against some wines. And while it would be natural to go for a fairly acidic wine to cut the richness of the soup, since that richness comes from cream, which tends to linger in the mouth and affect the flavour of what follows, many crisply acidic wines end up losing their fruit and tasting overly sharp. What is required is freshness, certainly, but also enough fullness and depth of flavour not to be overwhelmed by the soup. Yet, if the wine is too rich, the vichyssoise can seem thin and tasteless.

New wines from Savoie

In the far east of France, just below the lake of Geneva, is the area of Savoie, whose wines have only recently begun to be seen around in any quantities. It is a pity we have had to wait so long, for there are some unusual and very pleasant tastes to be had. **Vin de Savoie Apremont Jean Cavaillé** is a case in point. It comes from around the town of Apremont in the foothills of the Savoie mountains and is made from the Jacquère grape, a rare variety found more or less only in Savoie. The wine has a very pretty tangerines and mountain flowers bouquet and a crisp but quite full floral and green fruits taste – just right for partnering vichyssoise. It is strongly flavoured enough not to be overwhelmed by the soup (but not so strong that it does the overwhelming), crisp enough to be refreshing, but rounded enough to balance the creaminess. In short it is a great partner, superb as an aperitif too and a delicious new taste.

Dry wines from Germany

German wines can be some of the most delightful to drink but often are not as suitable as they might be for accompanying foods as their sweetness is too intrusive. Recently, though, there has been a major movement in Germany to produce dry wines — usually called Trocken – and they have become very popular in their homeland. Abroad, we have been more wary of them, especially since without the characteristic touch of sweetness German wines often seem to be 'missing' something in their flavour. But dry German wines really come into their own when partnering meals. In particular, **Louis Guntrum 1987 Sylvaner Dry** from the Rheinhessen region of Germany is ideal with vichyssoise. Its gently, steely, florally-perfumed bouquet almost gives the impression that the wine is going to be too perfumed and aromatic for the soup. But because it is dry, and firmly steely in taste, it is just right. The taste of the soup sings through cleanly, tasting almost better than without the wine, while the wine gains additional weight and balance from the soup and seems fuller and more rounded.

Just across the border into France from the German wine regions lies Alsace and Alsace wines are somewhat similar to German wines too, although they are usually fuller in body and dry. So it is hardly surprising that a wine from Alsace is also a good vichyssoise partner. **Alsace Pinot Blanc,** from the Cave Vinicole Eguisheim, has a gentle, creamy but aromatic bouquet and if fully-flavoured, quite powerfully so, with creamy, appley, slightly marzipan-like fruit. It is almost too weighty for the vichyssoise – a more powerful Alsace wine might well be so – but it just manages to balance well, with the flavours of both soup and wine complementing each other nicely.

A fine touch of Fino

As soup is a liquid, the purpose of drinking wine with it is only for a contrast of flavour, not to refresh the mouth or as an aid against thirst. So only the tiniest sips of wine may be needed. This means that sherry, which can be quite strong, is well worth considering. It has to be a dry sherry, though, and there are few better than **Puerto Fino Superior Dry Sherry** from Burdon. Fino is the lightest, driest type of sherry, with a particular character from 'flor', a special type of yeast which forms a layer on the sherry while it is developing in its oak barrels in its homeland in and around the town of Jerez in south west Spain's Andalucia. The perfect aperitif, with its great elegance of flavour and its yeasty, salty tang, it would be perfect to take your starter of Puerto Fino to the table and continue to have a few sips of it with vichyssoise. Although the sherry has quite a lasting flavour, it partners the soup so well that the flavours of both the sherry and the soup are brilliantly enhanced, and the contrast between the two quite different flavours is enlivening. Even more convenient, Puerto Fino is available in half bottles too, the ideal quantity if you are dining *à deux*, or have only a few friends round for dinner, especially since it is far better not to keep a good fino sherry like this too long once it is opened. The fresher it is, the better. Keep the bottle unopened in the refrigerator until you are ready to serve, then serve it chilled - whether the soup is served cold or hot.

PERFECT PARTNERS

Vichyssoise

- ***Preparation: 15 minutes***
- ***Cooking: 40 minutes***

50g/2oz butter
1kg/2¼lb leeks, trimmed of green tops, cleaned and sliced
100g/4oz potatoes, peeled and sliced
1L/1¾pt chicken stock
salt and freshly ground black pepper
large pinch of freshly grated nutmeg
425ml/15fl oz single cream
2tbls snipped chives

- ***Serves 6-8***

1 Melt the butter in a saucepan over a low heat. Add the sliced leeks, cover and cook very gently for 15 minutes, stirring occasionally. Do not allow the leeks to brown.

2 Add the potatoes and chicken stock to the pan, cover and simmer for 20 minutes or until the potatoes are tender.

3 Purée the soup through a sieve or in a blender. Season and add nutmeg.

4 Leave the purée to cool, then stir in most of the cream and chill. When ready to serve, turn the soup into a chilled tureen, swirl the remaining cream over the surface and garnish with snipped chives.

Variations

Vichyssoise proper is always served chilled. The hot version, leek and potato soup, is also good. To serve hot, return the purée to the rinsed-out pan, stir in most of the cream and reheat gently.

ROBUST WINES FOR SMOKED FISH

PINOT GRIGIO

Choosing the wine to go with fish should be no problem. But how to complement the assertive flavours of smoked fish?

JUST BECAUSE A fish is smoked doesn't mean it begins to taste like all other smoked fish; just to think for a moment of the difference between smoked haddock and smoked salmon will show *how* different.

Nevertheless, several types of fish, like trout or mackerel – they could be the first course of a dinner party or the main course of a light supper or lunch – have certain common characteristics that make similar sorts of wines go well with them.

Earthy Mâcon

Apart from the positive smoky flavour derived from the process, smoking tends to accentuate the taste of the fish itself and to make the flavours more robust. So it is unlikely that a wine that will complement a fish in its unsmoked state will be the same ideal partner when the fish has been smoked. It should 'go' to a certain extent, but others will go considerably better. Take Chablis, for example, the ideal wine to serve with so many fish dishes; when the fish is smoked most accompanying Chablis wines just seem to lack something. But if, instead of Chablis, which comes from northern Burgundy, you try a wine like **Mâcon-Viré 1987 Domaine des Chazelles,** which is produced farther south in the same region, the match is far better.

Mâcon has basically similar characteristics to those of Chablis but, coming from a warmer climate, it has better body and – a particular quality of this wine – a slight touch of earthiness that is lost against the smokiness of the food.

Domaine des Chazelles is a particularly well-balanced, rounded, crisp and concentrated example of Mâcon and a most impressive wine. You can't go wrong next time you offer smoked trout or mackerel.

Bourgogne, robust but flowery

On similar lines, but bigger, more robust and therefore better suited as a main course or all-meal wine is **Bourgogne 1987 Domaine Sainte Claire** – from the same region, the same grape (Chardonnay) and the same vintage as Domaine des Chazelles but with an extra weight and butteriness. On the nose it gives a hint of orange, on the palate plenty of concentrated flavour with a most

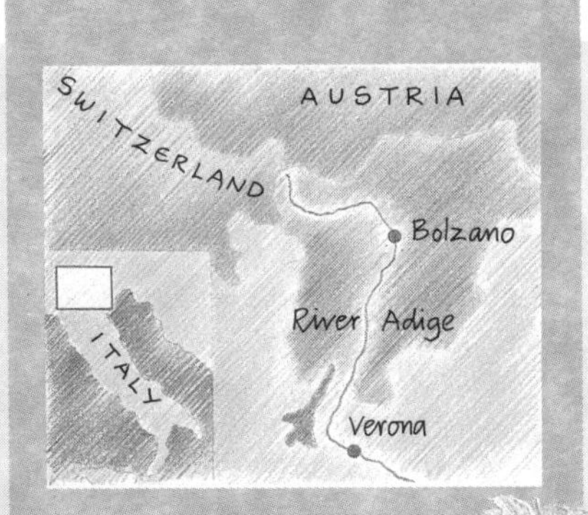

BEST BUYS

Tiefenbrunner Pinot Grigio
***Taste guide:* dry**

Mâcon-Viré 1987 Domaine des Chazelles
***Taste guide:* dry**

Bourgogne 1987 Domaine Sainte Claire
***Taste guide:* dry**

La Jaubertie 1988 Bergerac Sec
***Taste guide:* very dry**

Glen Ellen 1987 California Chardonnay Proprietor's Reserve
***Taste guide:* dry**

attractive flowery touch; giving elegance to prevent it overpowering fish like trout.

Oily but nice

An absolutely brilliant match with smoked fish is made by **Pinot Grigio 1988 Tiefenbrunner.** If the name confuses you it is because the wine is Italian but comes from the country's extreme north east, the zone called Alto Adige, where many of the population are German and German-speaking, as is Herr Tiefenbrunner, its producer. The zone is called Süd Tirol, and both the German and Italian variations on its name appear on the wine's label. The Pinot Grigio grape grows especially well in this part of Italy.

Refreshingly acid

From south-western France comes **La Jaubertie 1988 Bergerac Sec.** Bergerac is a biggish region near Bordeaux that produces good red, dry white and sweet white wines, and La Jaubertie is one of the best of them, all the better because it is made by an Englishman, Henry Ryman, who settled in Bergerac some years ago. This dry white is a blend of grape varieties, mainly Sémillon and Sauvignon; it gets a character of ripe gooseberries from the latter and a soft, fleshy, lanolin-like quality from the former.

The fruit is quite flowery, and a lovely refreshing acidity makes the wine an excellent partner for the oilier smoked fish like mackerel, cutting right through it and leaving the mouth perfectly clean. Its lively fruit means that it also works well with less oily smoked fish.

Light from the USA

A successful but rather surprising match is **Glen Ellen 1987 California Chardonnay Proprietor's Reserve** – surprising because the rich, oaky, melon-and-pineapple character of California Chardonnays usually renders them too powerful when combined with fish to allow all its restrained flavours to come through. This wine, though, is made in a light style, with no overbearing flavours, and its delightful lemony tone doubtless explains its remarkably happy marriage with fish.

The wine is quite full flavoured and goes well with smoked fish, especially when the fish is combined with mild-tasting ingredients, or a sauce. The cucumber fish mousse, below, is delicious served with a cold tomato sauce: cook 1 chopped onion and 1 crushed garlic clove in olive oil until soft, add a 400g/14oz can chopped tomatoes and cook for 3 minutes. Process until smooth, then cool.

PERFECT PARTNERS

Cucumber fish mousse

- ***Preparation: 25 minutes, plus 1 hour chilling***
- ***Cooking: 40 minutes***

1 cucumber, peeled and cut into 3 sections, ends removed
225g/8oz smoked trout, mackerel or buckling, filleted
2 medium-sized eggs, separated
salt and white pepper
4tbls thick cream
2tbls tomato ketchup
1tbls lemon juice
about 300ml/½pt tomato sauce (see above)

For the garnish
raw carrot 'match sticks'
raw turnip 'match sticks'
cooked green beans
vinaigrette dressing

- ***Serves 4***

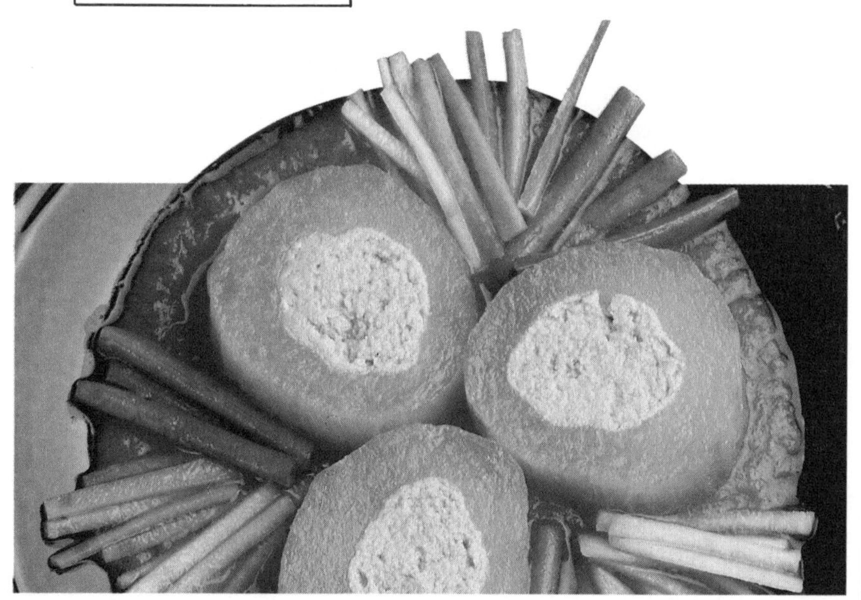

1 Place the smoked fish in a blender with the egg yolks and process until smooth. Add salt and pepper to taste and the thick cream and blend again. Add one egg white and blend until the mixture is light and fluffy. Add the tomato ketchup and lemon juice and blend until well mixed. Chill for 1 hour.

2 Using an apple corer, remove the cucumber seeds and inside of the cucumber sections, leaving a shell of cucumber 5mm/¼in thick all round.

3 Place each cucumber section on foil, fill with chilled mousse and wrap in the foil. Put the foil packages in a saucepan, cover with water and simmer for 30 minutes. Remove from the pan and allow to cool. Refrigerate until needed.

4 Meanwhile make the tomato coulis and leave to cool. Cook the carrot and turnip match sticks in boiling salted water for 5 minutes. Trim the green beans to the same length as the carrot and turnip strips. Marinate the vegetables separately in vinaigrette dressing for at least 1 hour.

5 To serve, divide the tomato sauce among 4 chilled side plates. Slice each cucumber section into 4 rounds, and place 3 slices on each plate. Garnish with the marinated vegetables.

PERFECT PRAWNS

QUINCY 1988

The delicacy of prawns is enhanced by white wines of subtle lightness

PRAWNS HAVE COME a long way from the ubiquitous prawn cocktail. The easy availability of frozen prawns has made them a popular freezer item, despite their expense, and has removed any fears about their freshness. Prawns now crop up in all sorts of recipes, from just a garnish, to a major, or even the sole constituent of a dish.

When thinking about what wine to serve with prawns, there are several simple rules to be borne in mind. Firstly, stick to white. Although it is getting more and more fashionable to drink red wine with fish-based dishes, prawns are much better with white wines. Secondly, prawns are quite delicate in flavour, so ignore full-bodied wines and go for something quite light and elegant. Even then, if the wine's fruit is too rich or its style too broad or nutty, it still won't go well with prawns. Good acidity is an important advantage and the drier the wine the better. To help you choose, a simple rule of thumb is to go for a wine that comes from near a coast where prawns are plentiful. It is another example of the almost infallible rule that foods and wines from the same origin will go well together. However, prawns do not only match coastal wines.

Look along the Loire

A superb match is **Quincy 1988 Domaine de Maison Blanche**. Quincy comes from the Loire valley of north-west France. At the mouth of the valley is the area where Muscadet is grown, a wine famous for its partnership with seafood. Although Quincy comes from further inland, the maritime influence of the north Atlantic is still present enough to make Quincy a good bet for drinking with prawns. Even more encouraging, Quincy is made from the Sauvignon grape, a grape renowned for its great match with all sorts of fish dishes and for its good acidity. Domaine de Maison Blanche is full of refreshing, herbaceous and gooseberry-like aromas. It is a beautifully well-made wine; clean, crisp, fresh, with delicious but not too incisive gooseberry-like fruit. It provides an excellently refreshing contrast to the prawns, with much the same effect as squeezing lemon over them.

The mouth of the Loire river does not provide only Muscadet. Another

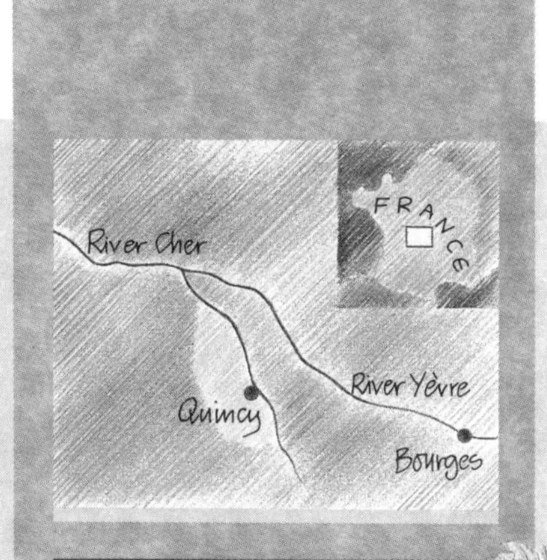

BEST BUYS

Quincy 1988 Domaine de Maison Blanche
Taste guide: very dry

Gros Plant du Pays Nantais 1988 Guy Bossard
Taste guide: very dry

Frascati Superiore 1988 Azienda Vinicola SAITA
Taste guide: dry

Verdicchio dei Castelli di Jesi Classico 1988 Colle del Sole
Taste guide: dry

Fino Sherry Bodegas Garcia
Taste guide: very dry

grape, the Gros Plant, grows there too, producing a wine called **Gros Plant du Pays Nantais**. Since wines from Gros Plant are light with refreshing acidity and since they come from a major seafood-eating area it is hardly surprising that they go perfectly with prawns. Wines from the Gros Plant grape have not been all that common outside of their growing region, but are thankfully becoming more so. The one to go for is 1988 bottled by Guy Bossard. This wine is special not only because it is produced organically, with no chemical weedkillers, fertilisers or insecticides used on the soil but also because it has a purity of taste that allows the limey, rounded, lively-fresh flavours of the grape to shine through. It will thoroughly enliven any plate of prawns.

Italian partners

Some of the best prawns come from around the Mediterranean so wines from the Mediterranean are often ideal prawn partners, Frascati, for example. It comes from the hills behind Rome, not far from the sea, and is drunk by Romans both in the city and along the coast. Try **Frascati Superiore 1988 Azienda Vinicola SAITA**. Made from a blend of Malvasia and Trebbiano grapes, it is attractively flavoured, with plenty of crispness, lemony and appley fruit and just a touch of light nuttiness. If it were any fuller-bodied it would be too much of a good thing for the delicacy of the prawns but as it is it provides a fascinating counterpoise to their flavours.

From the other side of Italy, the Adriatic coast, where seafood-eating is even more prevalent, there is the Verdicchio grape. It is one of Italy's best grapes for seafood, and thus for prawns. The wines it makes have enough weight to be interesting in their own right, without any risk of overwhelming even the most subtleof seafood tastes. One of the best of them is **Verdicchio dei Castelli di Jesi Classico 1988 Colle del Sole**. Classico means that the grapes come from the central heartland of the zone, historically the best for the wine and Colle del Sole is a particularly prized vineyard where the grapes grow. With that pedigree it is hardly surprising that it makes an elegant match for prawns.

Sherry and tapas

In Spain there is no shortage of prawns either, particularly from Andalucia in the south west, the region at the gateway to the Mediterranean. In Spain, it is common to eat prawns and other seafood as *tapas*, small plates of delicious nibbles that accompany drinks at bars and that can quite easily become a meal in themselves. The most usual drink is Fino Sherry, bone dry and much lighter and crisper than most types of sherry. Although you might think that sherry would be too big and weighty for prawns, Fino Sherry is an exception and its clean, salty style provides the perfect balance. To see this remarkable match try a bottle of **Fino Sherry** bottled by Bodegas Garcia with a plate of unadorned prawns. For a simple, tasty and most unusual starter, it is hard to beat.

If you prefer to serve prawns hot, then either use cooked, peeled prawns and dip them in a little spicy batter for deep-frying (see below), or use whole prawns still in their shell and simply grill or barbecue them. Marinate them first in olive oil, a little lemon juice and a finely crushed clove or two of garlic, if you like, then cook for just a few minutes until hot.

PERFECT PARTNERS

Deep-fried prawns

• ***Preparation: 15 minutes***

• ***Cooking: 10 minutes***

50g/2oz cornflour
1tsp turmeric
½tsp chilli powder
1tbls ground cumin
salt and pepper
oil for deep frying
225g/8oz cooked peeled prawns, defrosted if frozen
lemon wedges, to serve

• ***Serves 4*** (¶) (££)

1 Mix together the cornflour, turmeric, chilli powder and cumin with salt and pepper to taste.

2 Half fill a deep fat fryer with oil and heat to 180C/350F. Dry the prawns on absorbent paper, coat them individually in the seasoned flour and fry for 1-2 minutes, until crisp and golden.

3 Remove the prawns with a slotted spoon and drain them on absorbent paper. Serve hot, with wedges of lemon.

WINES FOR FISH AND CHIPS

VINHO VERDE

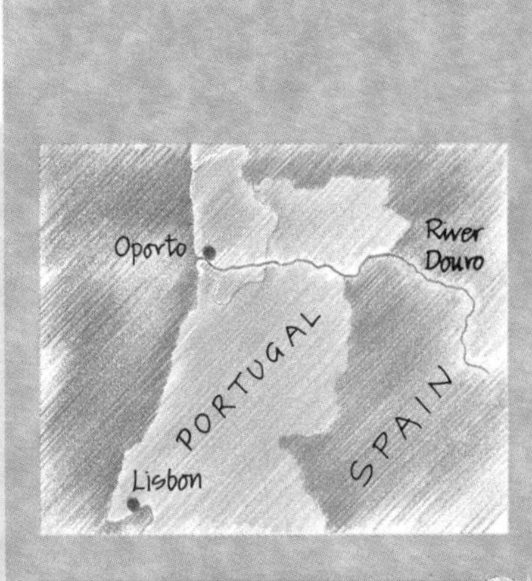

BEST BUYS

Vinho Verde Borges & Irmão
***Taste guide:* medium-dry**

Trebbiano d'Abruzzo 1987 Cantina Tollo
***Taste guide:* dry**

Sauvignon
***Taste guide:* very dry**

Alsace, Cave Vinicole de Bennwihr
***Taste guide:* dry**

Bourgogne Chardonnay 1986 André Ropiteau
***Taste guide:* dry**

It may sound like drinking tea with caviar, but fish and chips and wine can complement each other brilliantly

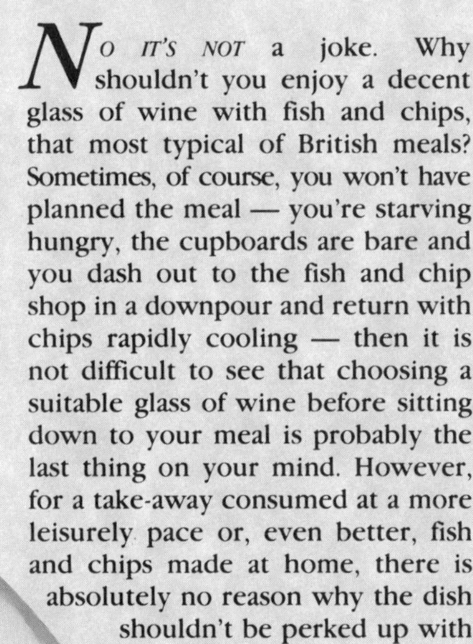

NO IT'S NOT a joke. Why shouldn't you enjoy a decent glass of wine with fish and chips, that most typical of British meals? Sometimes, of course, you won't have planned the meal — you're starving hungry, the cupboards are bare and you dash out to the fish and chip shop in a downpour and return with chips rapidly cooling — then it is not difficult to see that choosing a suitable glass of wine before sitting down to your meal is probably the last thing on your mind. However, for a take-away consumed at a more leisurely pace or, even better, fish and chips made at home, there is absolutely no reason why the dish shouldn't be perked up with wine. It is not that easy, though, to find a good match. You need wine with a reasonable amount of flavour to stand up to the batter or breadcrumb coating of the fish and the fried exterior of the chips. On the other hand, it can't be too powerful, otherwise it will completely overwhelm the delicate flavour of the fish. One word of warning: if you like your chips drenched in vinegar you won't be able to taste much of *whatever* wine you serve.

A good catch

Often, wines that are fairly ordinary on their own really excel when partnered with the right foods. Fish and chips improves a fair number of white wines. **Trebbiano d'Abruzzo 1987 Cantina Tollo** comes from Italy's east (Adriatic) side, towards the south. It has a firm nose of thyme, and a taste that is dry, clean, attractive, not too acidic, but a touch earthy. Nice enough on its own, it improves out of all recognition with fish and chips. It has just the right weight to complement the delicacy of the fish and its firm, low acid style blends wonderfully with the other ingredients.

In the same way,. **Vinho Verde Borges & Irmão** tastes about three times as impressive with fish and chips as on its own. The way the partnership works, though, is different. Vinho Verde is very high in acidity and this acts as a condiment to the dish, rather like putting lemon on fish, and perks it up. It wouldn't necessarily be the same with all Vinho Verde. This one has only the slightest of fizz and just a touch of sweetness. Vinho Verde can be anything from bone dry to really quite sweet. The driest are usually only drunk in Portugal, its homeland; most bottles sold in Britain are well on the sweet side. It is vital to buy from somewhere with a fast stock turnover – Vinho Verde tires very quickly, and without the lovely, zippy freshness of the Borges & Irmão the delicious partnership of wine and food would be lost.

Netting the best

Another zingy fresh wine that adds life to the food is **Sauvignon**. Sauvignon is a grape variety most at home in northern France, particularly around the Loire valley, where it is often called Vin de Pays du Jardin de la France. It has a delightful gooseberry character. It is the gooseberryishness that enlivens the fish and cuts through any oiliness. The simpler the wine is, the less intensely flavoured it is, therefore the more it complements the dish and the less it fights to dominate it. Hence it is best just to go for the simply titled French Sauvignon.

Still waters

The wines of Alsace, from eastern France near the German border, are strangely underrated. They deserve to be much better known as they combine everything that makes a white wine appealing: they are elegant and aromatic, dry but not searingly dry, they have masses of fruit, good weight, and are beautifully crisp. They are also remarkably good partners for food: they have enough mouth-watering character to pep up the taste buds and enhance the taste of whatever food follows; they act as a counterpoint to the food, rather as a good relish does; and their flavour is rarely adversely affected by whatever food they partner. Fish and chips is no exception. Again, a more straightforward wine, called simply Alsace, made from a mixture of grapes, is better with fish and chips than anything more complicated. An ideal example is **Alsace** from the Cave Vinicole de Bennwihr.

If you like a bit more body in a wine, it is helpful to choose one that has spent a little time ageing in oak casks, but from a grape variety like Chardonnay that gains complexity from oak and isn't overwhelmed by it. **Bourgogne Chardonnay 1986 André Ropiteau** is the perfect example. Rounded, rich and oaky-buttery on the nose, with a taste of appley freshness which joins the rounded, smooth butteriness - this wine would be great with a wide variety of dishes. Although, at first glance, a rich and oaky Chardonnay - especially one from France - might seem more than a little too weighty to serve with fish and chips, it isn't. Indeed the Chardonnay highlights the flavour of the fish with its crisp and tasty batter coating, and marries extremely well with the chips, while its own strong, yet fresh, appley taste stays virtually unchanged.

PERFECT PARTNERS

Fish and chips

- ***Preparation: 35 minutes***
- ***Cooking : 20 minutes***

100g/4oz self-raising flour
salt
groundnut oil, for deep-frying
4 × 175-225g/6-8oz pieces of cod or haddock, skinned
flour, for coating
1.1kg/2½lb potatoes, cut into 1cm/½in wide chips
lemon wedges and parsley sprigs, to garnish

- ***Serves 4***

1 Heat the oven to 100C/200F/gas low. Sift the flour and 1tsp salt into a large bowl. With a wooden spoon, gradually beat in 150ml/¼pt water, drawing the flour in slowly from the sides, to form a smooth batter.

2 Heat the oil in a heavy saucepan to 190C/375F. Pat the fish dry with absorbent paper, dust with flour to coat thoroughly, then dip the floured fish into the batter and deep-fry, in batches, for 4-6 minutes or until the fish is cooked and the batter is golden brown.

3 Drain the fish on several layers of absorbent paper, then line a baking sheet with some more, and place the fish on this in the oven to keep warm.

4 Rinse the chips under cold water and dry thoroughly with absorbent paper. Check that the oil is heated to 190C/375F, then fry the chips for 5-6 minutes or until golden brown. Drain on absorbent paper. Serve immediately with the fish, garnished with lemon wedges and parsley.

CHEAP AND CHEERFUL WHITES

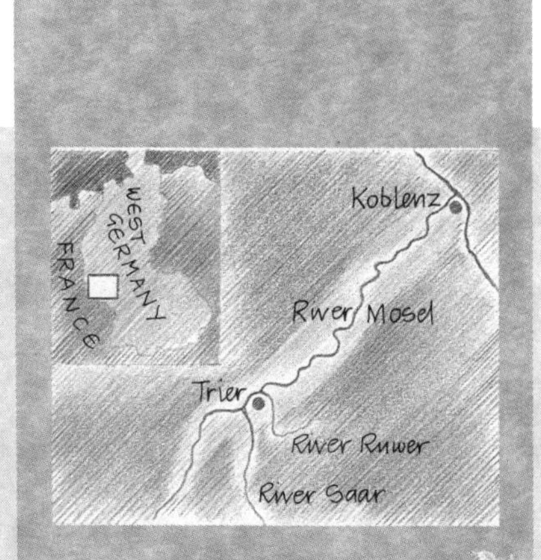

BEST BUYS

Riesling 1988
***Taste guide:* medium-dry**

Vin de Pays d'Oc Cépage Mauzac
***Taste guide:* very dry**

Vin de Pays de l'Hérault Union des Grands Chais
***Taste guide:* medium-dry**

Vin de Pays de l'Ardèche 1987 Les Vignerons d'Ardèche
***Taste guide:* medium-dry**

Cantina Sociale di Dolianova Carafe White
***Taste guide:* dry**

Zagarron 1988
***Taste guide:* dry**

RIESLING 1988

White vins de pays are easy on both palate and pocket

MOST OF THE time, when looking for something to drink, you may well not be too worried about distinguishing this or that particular characteristic of a wine – all you want is something pleasant, that will go down well enough and won't make you feel rotten the day after: something cheap and cheerful. It would be nice to think all you need do is search out the cheapest bottle you could find and the cheaper a wine is the more innocuous it is. Unfortunately, this is only true in some cases. In others cheap and nasty is a more apt description.

Vin de Table

For example, although it may be tempting to go for something French labelled simply as 'Vin de Table', the grapes that go into these wines are often blended to get a uniform taste. What can happen is that the companies making Vins de Table have to buy the cheapest grapes and wines and may get the left-overs people don't want. So, although some wines called just 'Vin de Table' are fair enough, others are pretty dismal.

Vin de Pays

It is far better, when looking for a French wine, to go for a Vin de Pays. These are also Vins de Table, but a special superior category of them because the grapes come from a specific part of the French countryside. They therefore have 'the taste of the country'. They are also usually inexpensive and very good value. An example is **Vin de Pays d'Oc Cépage Mauzac.** It comes from the south of France, around the area better known as the Languedoc. An added plus point to the wine is that it comes from just one grape variety, Mauzac, which is found only around the south of France and is ideally suited to producing wine in that particular climate and on the south's soils. Crisp, fresh, grassy and herby, it will slip down very easily.

Although most Vin de Pays are dry, you may prefer an off-dry or medium-dry wine. There are some Vin de Pays that will suit, like **Vin de Pays de L'Hérault** bottled by Union des Grands Chais. The Hérault is a département, also in the south, south and west of Montpelier. Fresh and herby-spicy, quite lightweight and with just a touch of sweetness, this Vin de Pays is fine for almost any occasion. You may prefer to look for a little extra weight and a little more sweetness. If so, try **Vin de Pays de L'Ardèche 1987** bottled by Les Vignerons d'Ardèche. The Ardèche is also a département, just a bit further north, by the west bank of the river Rhône. The wine smells slightly honeyed, and it tastes lively, rounded and biscuity-crisp. But, if you are looking for medium-dry wines, do check back labels carefully since there are wines from both Hérault and Ardèche which are dry. Check that the label has a bottle-shaped sweetness indicator bearing the number 3, 4 or 5 (1 or 2 means dry).

France is not the only country producing cheap and cheerful quaffers; most European countries do. Surprisingly, it is harder than you might imagine finding something fresh and lively from Germany, many wines are milky-soft and undistinguished. One exception is **Riesling 1988.** It comes from the Mosel-Saar-Ruwer region (which can produce some of Germany's best wines) although this is not given prominence on the label; the important thing is that the wine is made from the Riesling grape variety, which can make some deliciously fruity, floral wines. Crisp, grassy and flowery, light, elegant and medium-dry, this is a delight to drink without being a strain on the pocket.

From Italy, too, there are some great bargains, such as the wine bottled by the Cantina Sociale di Dolianova and shyly labelled simply **Carafe White**. It is far better than anything you would expect to see in a carafe, with its lively, green fruits and kernels bouquet and its dry, firm, rounded, crisp, grassy taste. It is a wine for any occasion and, like the Riesling, comes in a litre bottle so there is plenty for a large group without the expense of buying two bottles.

Spain used to be the country everyone looked to for providing plonk, however it tasted. Now there is still plenty of inexpensive wine, but much, much less that deserves the term 'plonk'. An ideal example of the result of the new skills that have been turning out vastly improved wines is **Zagarron 1988**. It is from La Mancha, the vast high plain of central Spain that is covered with acres and acres of vines. There is little rain and the land bakes by day and gets pretty chilly by night. So La Mancha needs special grape varieties that can cope with such a climate. Airen, which is used for Zagarron, is one. The wine is as clean and fresh as you could wish, with flavours including lemon, liquorice and pears; not just cheap and cheerful, but cheering too. Which, not surprisingly, is also an apt description of the food that goes well with these wines. Hearty casseroles, hot-pots, even a bowl of chilli, would go down well. Or, if you prefer fish, then choose a robust, flavourful dish such as cod and bacon au gratin, below.

PERFECT PARTNERS

Cod and bacon au gratin

- ***Preparation: 30 minutes***
- ***Cooking: 50 minutes***

4 fresh cod steaks, about 175g/6oz each
50-75g/2-3oz butter
8 thin slices streaky bacon, rind and bone removed
2 onions, finely chopped
350g/12oz potatoes, sliced
salt and pepper
½ bay leaf
2tbls chives or spring onion tops, finely chopped
2tbls finely chopped parsley
4tbls thick cream

- ***Serves 4***

1 Heat the oven to 180C/350F/gas 4. Heat 25g/1oz butter in a frying pan, add the bacon slices and fry lightly on both sides. Remove the bacon from the pan and keep warm. Add the finely chopped onions to the pan and sauté gently until transparent, adding an extra 15g/½oz butter.

2 Parboil the sliced potatoes in boiling salted water for 5 minutes, taking care that they do not become overcooked. Drain thoroughly and keep hot.

3 Butter a deep heatproof baking dish. Lay the ½ bay leaf on the bottom Cover with half of the parboiled potato slices, arranging in neat overlapping circles. Sprinkle with half the sautéed onions and 15ml/1tbls each of the finely chopped chives or spring onion tops and parsley.

4 Arrange 6 of the sautéed bacon slices on top and cover with the cod steaks. Sprinkle the steaks with the remaining chives or spring onion tops and parsley. Cover with the remaining sautéed bacon slices and onions, and finally with the remaining potato slices. Season generously with salt and freshly ground black pepper, and pour over the thick cream.

5 Dot with 15g/½oz butter and bake in the oven for 40 minutes, until the potatoes and fish are cooked through. Brush with melted butter and brown under the grill if liked.

VERSATILE WHITES

MUSCADET

Always cool and refreshing, dry white wines range in character from the delicate to the full-bodied

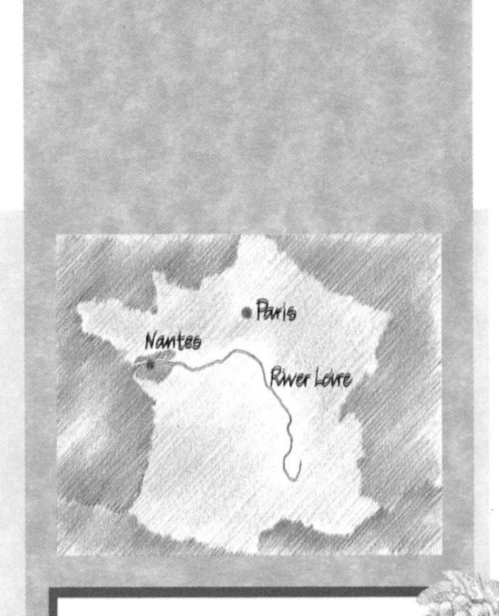

BEST BUYS

Château de la Galissonière, Muscadet de Sèvre et Maine 1986
Taste guide: ***very dry***
Serve with: ***fish and shellfish***

Campo Burgo white Rioja 1987
Taste guide: ***very dry***
Serve with: ***salads, picnic foods***

Colli di Catone Frascati Superiore 1987
Taste guide: ***dry***
Serve with: ***chicken, light pasta dishes, quiches***

Canterbury California Sauvignon Blanc 1986
Taste guide: ***dry***
Serve with: ***barbecues, spicy food***

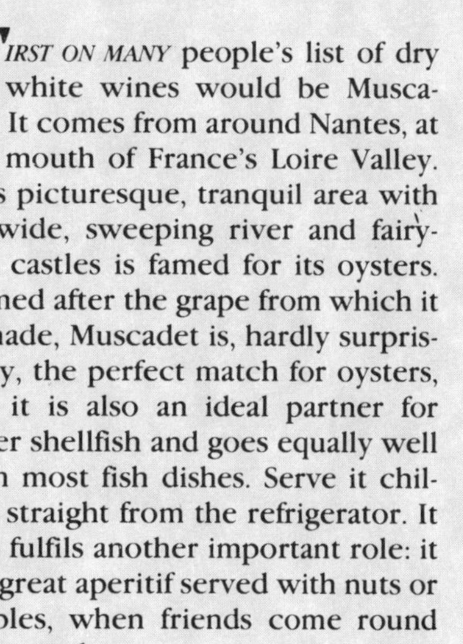

FIRST ON MANY people's list of dry white wines would be Muscadet. It comes from around Nantes, at the mouth of France's Loire Valley. This picturesque, tranquil area with its wide, sweeping river and fairy-tale castles is famed for its oysters. Named after the grape from which it is made, Muscadet is, hardly surprisingly, the perfect match for oysters, but it is also an ideal partner for other shellfish and goes equally well with most fish dishes. Serve it chilled, straight from the refrigerator. It also fulfils another important role: it is a great aperitif served with nuts or nibbles, when friends come round for a meal.

GLASSES FOR WHITE WINES

The most important thing is that the glass has a stem. To refresh, and to taste its best, white wine should be chilled. In summer it can warm up pretty quickly anyway, but a hot hand cupped round the bowl of a glass will speed the process considerably, and the wine will soon lose part of its appeal. Holding a glass by its stem avoids this problem.

Glasses are best if they are reasonably large — but not huge — so that a decent serving fills them no more than half full. They should also curve in towards the rim; this conserves the bouquet and directs it to your nostrils as you sip, adding to your pleasure. A straight-sided or slope-sided glass helps those glorious aromas escape into the air. What a waste!

The area around the rivers Sèvre and Maine produces particularly good Muscadet and it is from here that **Château de la Galissonière, Muscadet de Sèvre et Maine 1986** comes. A clear, bright straw colour, it is as crisp as biting into a good apple and as lively as a lemon sauce. Dry, light-bodied, fresh and thirst-quenching, best drunk chilled but not icy cold (like all white wines), there are few wines that can match it for easy drinking pleasure.

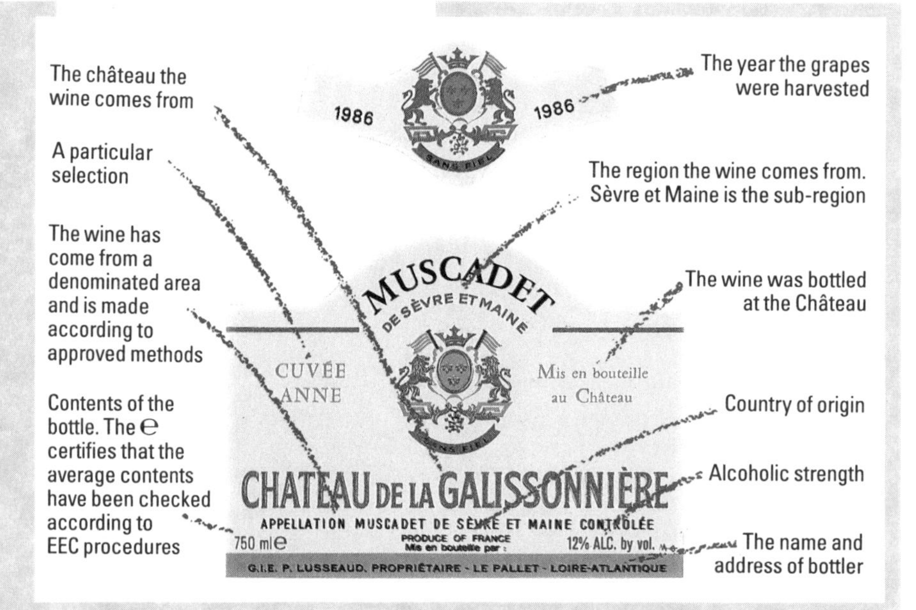

Other versatile whites

Campo Burgo white Rioja 1987 is one, though. Rioja, in northern Spain's Ebro valley, is much better known for ripe, oaky red wine. But there's plenty of white made too. Campo Burgo is especially appealing. Still full of youthful fruit, the mixture of zippy, citrus flavours and rounder, gentle apricot-like tastes instantly lifts the spirits.

For a lunch based on chicken or a quiche, Frascati fits the bill. Frascati is the wine of Rome, drunk freely in and around the seven hills, and is made from a mixture of peachy Malvasia and nutty Trebbiano grape varieties. One of the best is **Colli di Catone Frascati Superiore 1987.** Characteristically pale in colour, it has a heady smell redolent of cream, peaches and crushed walnuts. Dry and medium-bodied, the fruity, nutty taste is perfect when served with pasta or chicken.

White wine can sometimes be quite full-bodied. **Canterbury California Sauvignon Blanc 1986** is a good example. Imagine a fruit salad with plenty of gooseberries turned into wine. It's dry but well-rounded; forceful enough to stand up to red meat or spicy foods, yet not too strong for fish.

PERFECT PARTNERS

Cod in white wine sauce

• ***Preparation: 25 minutes***

• ***Cooking: 55 minutes***

75g/3oz butter
1 small onion, chopped
900g/2lb cod fillet, skinned and cut into serving pieces
1 carrot, in julienne strips
2 celery sticks, in julienne strips
2 small leeks, in julienne strips
300ml/½pt dry white wine
150ml/¼pt fish or chicken stock
25g/1oz flour
150ml/¼pt single cream
juice of ½ lemon
salt and white pepper

• ***Serves 4–6***

1 Use 25g/1oz of the butter to grease a large, shallow ovenproof dish and a piece of foil to cover the dish.

2 Cover the base of the dish with the onion. Place the fish on the onion. Heat the oven to 190C/375F/gas 5.

3 Melt 25g/1oz butter in a medium-sized pan over low heat. Add the julienne vegetables, stir, cover and cook for 8–10 minutes or until softened.

4 Spoon the vegetables over the fish, then pour in the wine and stock. Cover with the foil, then bake for 25–30 minutes.

5 Drain the liquid into a large pan; keep the fish and vegetables warm. Boil fast to reduce to 300ml/½pt. Cream together the remaining butter and the flour; beat into the liquid over low heat, piece by piece. Cook for 5 minutes or until thickened. Remove from the heat and stir in the cream.

6 Replace over heat and bring slowly to just below boiling. Add the lemon juice and season. Drain fish and vegetables; put on a platter and pour the sauce around.

SPIRITED SPARKLERS

IRON HORSE

Celebrations bring champagne to mind, but nowadays a good-quality sparkling wine from California or Australia may be a better choice. Indeed, you may prefer their richer taste

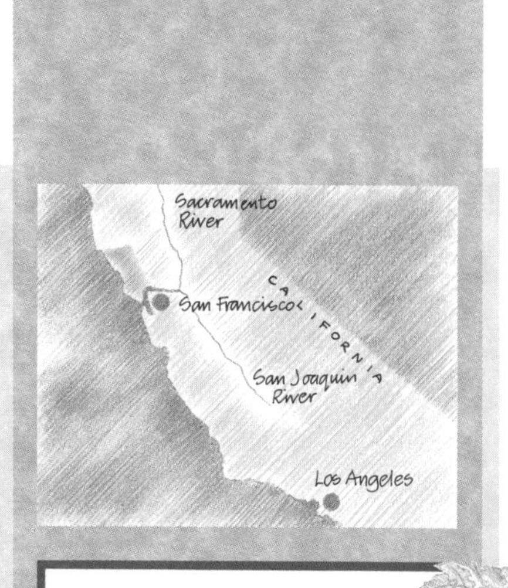

BEST BUYS

Iron Horse Brut 1984
***Taste guide:* dry**

Brut Sparkling Australian White Wine
***Taste guide:* dry**

Seaview Grande Cuvée
***Taste guide:* dry**

Seaview Brut
***Taste guide:* dry**

SPARKLING WINE IS always associated with celebrations. There is something about wine with bubbles that creates a mood of festivity even before a drop has been swallowed, and there are few people, with something fizzing gently in their glass, who can resist finding a reason to toast something or somebody. For many celebrations, champagne is the obvious choice. But there are several reasons why it may not be the right choice. First, and most important, there's the cost — champagne isn't cheap. Then there's the

OPENING THE BOTTLE

Opening sparkling wine bottles can be a perilous operation if not done with care. The pressure in a bottle can be as much as that in the tyre of a double-decker bus! So shaking the bottle and/or deliberately 'popping' the cork is not only wasteful but dangerously foolhardy. The correct way saves walls, ceilings, vases, television sets and eyes from damage.

First remove the foil. Next, untwist and remove the wire, keeping your thumb hard on the cork just in case the bottle is very lively and it starts to shift straight away. Usually, though, the cork is tightly wedged in the neck of the bottle.

Grip the cork with one hand (with the wire removed your hand won't be cut) and the bottle with the other. Put a cloth round it if it is wet and slippery. Slowly twist the bottle — not the cork, unless you want to break it in half.

Starting may be quite tough, but once the cork begins to move it can jump out all too easily. So keep twisting the bottle gently, and control the cork by pressing down with your hand so that it eases out slowly. The aim is to release the gas pressure slowly, with a hiss, not a pop. To achieve this, once the cork is almost out, tilt it slightly from the bottle so that the gas comes out at one side rather than from all sides at once. Once the pressure is released you can remove the cork completely from the bottle.

Sparkling wine should be served cool. This also reduces the pressure and therefore the risk of an explosion. If, though, the cork is too tight to budge, you should hold the neck of the bottle for a short time to warm it slightly and so increase the pressure. Or, if you have a lot of bottles to open, you could invest in champagne tongs. These look like nutcrackers and grip the cork to increase your leverage on it and can make the whole operation a lot less hard work.

taste; glorious as good champagne can be, you may just not like it. But there are many other sparkling flavours worth enjoying.

For these reasons, or just for something refreshingly different, it is well worth trying sparkling wines from the New World (mainly Australia and California). Most of them actually smell and taste quite similar to still wines from the same areas, but with bubbles. And, since the wine districts of these countries are quite warm, the grapes used are often riper and consequently the wines tend to richness and avoid sharpness.

Quality bubbles

Brut Sparkling Australian White Wine from Penfolds is one good example. A bright, yellow-straw colour, it has fine, even, long-lasting bubbles — a good sign of quality. Its bouquet is honeyed and gently herby, while it tastes rounded, ripe, just off-dry and, like its bouquet, a little herby. The 'feel' of the sparkle is good — it neither batters the mouth like a fizzy drink nor froths wildly. Another quality pointer is that the flavour lingers in the mouth after the wine has been swallowed.

Alternatively, again from Australia, there is **Seaview Sparkling Wine** which comes in two versions, **Brut** and **Grande Cuvée.** Both are made by the *méthode champenoise,* the classic method used in Champagne which is long and labour-intensive but which produces the finest sparkling wines. Seaview Brut is a pale straw colour with tiny, slowly rising but continuous bubbles. A sniff gives a light, dry and biscuity bouquet, and it tastes a bit like ripe yellow plums with fresh lime juice. The fizz is gentle on the tongue and the after-taste is clean. Seaview Grande Cuvée is just a little creamier and a touch less dry.

Extra-special occasions

If it is a very special occasion, California can provide the answer. **Iron Horse Brut 1984** from California's Sonoma County Green Valley is pricey but magnificent - a real treat. Just as with the classiest of wines, the bouquet almost stops you taking a mouthful; it's rich and yeasty and full of all sorts of fruits and flowers. Firm, sparkling bubbles, which keep on rising in the glass, the flavour of ripe, buttery, peachy fruit and a lingering, wonderful and delicious aftertaste all combine to ensure that your special celebration will be a real success.

PERFECT PARTNERS

Smoked fish rolls

● ***Preparation: 30 minutes***

225g/8oz smoked trout fillet
4tsp horseradish sauce
6tbls double cream
large pinch of cayenne pepper
lemon juice
2 bunches of watercress
12 thin slices of smoked salmon (about 400g/14oz)
lemon twists, to garnish
thins slices of buttered brown bread, to serve

● ***Serves 6***

1 Skin the smoked trout and flake into a large bowl. Add the horseradish sauce and cream and mix well with a fork until fairly smooth and creamy (don't blend it). Add the cayenne pepper and 1tsp lemon juice and stir until well mixed.

2 Pluck the leaves from one bunch of watercress. Finely chop 2tbls and add to the fish mixture. Pluck sprigs from the remaining watercress and reserve these for the garnish.

3 Put about 1½tbls smoked trout mixture at one end of each slice of salmon. With your fingertips, pat the mixture into a cigar shape, and then roll up the smoked salmon slices lengthways.

4 Lay the rolls on a serving platter, surround with the sprigs of watercress and garnish with lemon twists. Serve the smoked fish rolls with thinly sliced buttered brown bread cut into triangles, and squeeze some lemon juice over the rolls just before serving.

WINES TO IMPRESS

JALOUSIE GASCOGNE

Make Sunday lunch entertaining and an opportunity to sample a selection of highly drinkable white wines

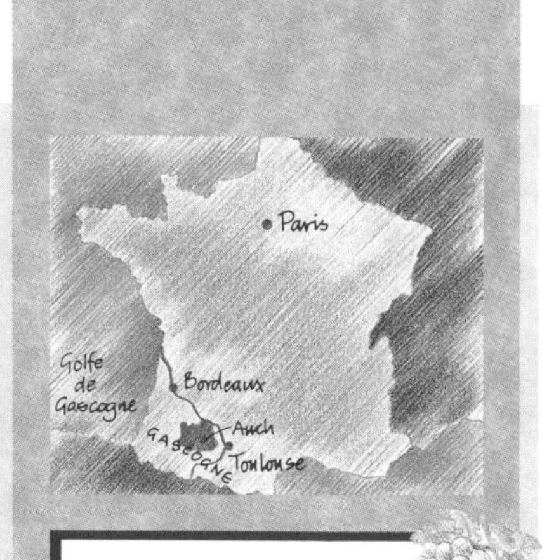

BEST BUYS

Domaine de la Jalousie 1986 Vin de Pays des Côtes de Gascogne Cuvée Bois
Taste guide: **very dry**
Serve with: **pâtés and terrines**

Sancerre, 1987 Domaine des Grosses Pierres
Taste guide: **very dry**
Serve with: **prawns, seafood and light fish**

Chablis Premier Cru Vaulignot 1985
Taste guide: **dry**
Serve with: **fish, avocado and asparagus**

Orlando Chardonnay 1987
Taste guide: **dry**
Serve with: **fried or rich food**

SUNDAY LUNCH FOR family and friends is an occasion to create a favourable impression – not only should you feel confident about the menu, but also about the wines. Wines can impress in several ways, from the classiness of a good, expensive bottle, to the lively, up-front appeal of a medium-priced, good-value buy; even a smart label and name can have a profound effect.

Specially blended

Take **Domaine de la Jalousie 1986 Vin de Pays des Côtes de Gascogne Cuvée Bois** which comes from Gascony in south western France. This is the region of Armagnac production and until recently known for little else. Then it was realized that the Colombard grape grown in the region could, when carefully made and blended with some Ugni Blanc to tone it down a little, produce a zingy-fresh, appley. delightfully drinkable wine.

At Domaine de la Jalousie it was decided to go one stage further and mature the wine in oak barrels. The wood has toned down the overt fruitiness of the wine, softened it and added an intriguing twist to the flavour.

Those who sniff wine seriously will notice it smelling mainly of oak, but in the mouth the fruit and the wood coalesce into a medium-weight, stylish taste. Use of oak barrels is usually the preserve of more expensive wines and a little oak on the taste often indicates something quite classy; the wine tastes elegant and harmonious; and the label is smart – all in all an impressive combination.

The young ones

One name to highlight is **Sancerre** and **1987 Domaine des Grosses Pierres** shows all the refinement of this well-known name. Sancerre comes from France's central vineyards along the Loire valley and, like Pouilly Fumé, its neighbour across the river, is made from the Sauvignon grape.

Sauvignon is characterized by its bouquet and taste of gooseberries which in a good Sancerre is more mute and musky, giving a gentler flavour but a more interesting one. As its style comes from its verdant freshness, Sancerre is best when reasonably young, from the summer following the vintage until a year or so later. Domaine des Grosses Pierres satisfies all these criteria and also avoids the excessive acidity that can mar some wines. It is another wine that looks and tastes right.

Chablis chic

Another name to consider, indeed *the* white wine name of all time, is Chablis which, unfortunately, has been rather expensive in recent years. The reason for this is that it comes from a northerly region of France, not far from Paris, where the weather is not as reliable as it is further south, and the fate of the grape harvest is always delicately poised between success and failure.

The weather also affects the flavour of Chablis from one year to another. For example, the 1983 vintages were big, rich wines, almost sweet; while those of 1984 were steely and sharp.

The worst risk to the vines comes from late frost which may damage its delicate buds. Despite measures – both simple and sophisticated – to avert this, the crop can still be decimated. Frost and poor summers between them can easily mean a shortage of Chablis, pushing prices up, which is just what happened recently.

If you decide to serve Chablis, it's worth considering a really good one like **Chablis Premier Cru Vaulignot 1985.** It is from one of the better Chablis vineyards, designated Premier Cru. (Chablis comes in four grades: Petit Chablis, Chablis, Premier Cru and Grand Cru. The first and last grades are only made in small amounts.)

Chablis is made from the famed Chardonnay grape. Often richly buttery, from the cooler climes of Chablis, the grape is creamier and, as with the Vaulignot, herby. It is the sort of wine to sip and savour. Quite full-bodied and fairly strong, it has a tremendous concentration of rounded, balanced, creamy-butteriness that lingers and soothes.

Australian choice

A clever buy to set lunch off with a swing is **Orlando Chardonnay 1987** from south eastern Australia. It is from the same grape as Chablis but grown in the warmer, much more reliable climate of Australia which gives it extra ripeness, weight and fleshiness.

Like most Australian Chardonnay, this wine is quite deep in colour and smells a little like butterscotch. It has also been matured for some time in oak which adds even more intricacy to the already fairly complex flavour. Full-bodied, but silky smooth and delightfully clean tasting, this delicious wine will be certain to make its mark with your guests.

Butterfly prawns

Only large prawns are suitable for this recipe, otherwise they do not hold their 'butterfly' shape. Serve this dish with crusty white or brown bread as an unusual and delicious starter. Your guests will enjoy eating the prawns in their fingers, as the tails are left intact to form natural 'handles'

- ***Preparation: 20 minutes***
- ***Cooking: 10 minutes***

450g/1lb cooked large prawns, in their shells
75g/3oz unsalted butter
1 garlic clove, halved
¼tsp ground ginger
1tbls light soy sauce
1tbls lemon juice
1tbls dry sherry
1tsp clear honey
lemon wedges, to garnish

- ***Serves 4***

1 Remove the head and body shell from the prawns, but leave the tail attached. Cut along the underside of each prawn, without cutting right through. Open out like a book but without breaking them.

2 Heat the butter in a large frying pan, add the garlic and ginger and mix well.

3 Add the prawns, cut side up, and fry until light golden, then turn and fry the other side. Transfer to a hot serving dish and then keep the prawns warm.

4 Add the soy sauce, lemon juice, sherry and honey to the pan, bring to the boil, then pour over the prawns or serve the sauce separately. Garnish with lemon wedges.

SPECIAL TREATS

TOKAY-PINOT GRIS

Choose a wine with a powerful personality to bring out the best in smoked salmon

WHEN YOU DECIDE to splash out and have smoked salmon you have a choice with the accompanying wine. You either decide 'in for a penny in for a pound' and decide to make a real occasion of it with a really smart wine or you decide to recoup a bit of the extravagance on the salmon with an inexpensive wine. Sadly, you can't go for a true bargain basement wine – not if you want to make the most of the smoked salmon, that is. For it can really show up the faults of a wine and make something that was pleasant enough on its own seem really nasty. But as the value of the wine in a bottle increases hugely for only a modest increase in cost (because the costs of shipping, duty and so on are fixed charges) you can find any number of not-too-expensive wines that are stylish enough to be salmon partners.

It takes a pretty powerful wine to squash the assertive smoky salmon flavours, so that's not the major problem in finding a good match. You need something, with complementary flavours which isn't affected by the taste of the salmon.

Whites from Alsace

One of the best choices is wine from Alsace, especially from the fatter, richer grape varieties like Pinot Gris or Gewürztraminer. There are several grape varieties grown in the region and they are nearly always vinified into wine separately with the name of the variety shown on the label. You won't go far wrong with **Alsace Tokay-Pinot Gris** from the Cave Vinicole de Turckheim. Tokay is another name for Pinot Gris and is the one in common use in Alsace. But there is also a Hungarian wine (from a different grape) called Tokai so to avoid confusion, the Common Market decided that if Tokay is used on labels of Alsace wine it has to be made clear it is the Pinot Gris, not the Hungarian type – hence the hyphenated name. That aside, the wine, either from the 1987 or 1988 vintages, is ideal with smoked salmon. It has a rich, nutty-spicy bouquet and a full, nut-kernels and spice, rich flavour which seems to have as mouth-coating a texture as the salmon itself. It is just the right weight to balance the salmon and is a perfect foil for it.

BEST BUYS

Alsace Tokay-Pinot Gris Turckheim Cave Vinicole de Turckheim 1988 or 1987
***Taste guide:* dry**

Alsace Gewürztraminer C.V.T. 1987
***Taste guide:* dry**

Casal di Serra 1987 Verdicchio dei Castelli di Jesi
***Taste guide:* dry**

Chablis Premier Cru Les Vaillons 1986 Alain Combard
***Taste guide:* dry**

Pouilly Fumé 1988 Les Griottes Michel Bailly
***Taste guide:* very dry**

If you prefer the more lychee-like spiciness of Gewürztraminer, with its broader, stronger flavours, go for **Alsace Gewürztraminer 1987** from C.V.T. Some Gewürztraminer is so powerful that it might well be too much for the salmon, however strong its taste. This wine, though, has more fruit than attack and is almost tailor-made for drinking with smoked salmon.

Super Verdicchio

Although it is normal to choose wines from the classic areas for classic foods, picking something a little more unusual can be very rewarding. **Casal di Serra 1987 Verdicchio dei Castelli di Jesi** is a case in point. Verdicchio is normally a light-bodied, fresh wine with good acidity from the Marche region on central Italy's Adriatic side, great with white fish like sole and plaice or with sea food. But Casal di Serra comes from a single vineyard on an especially good site and is made with particular care from low-yielding, fully-ripe grapes. The concentration and ripeness give a mellow biscuitiness that is rarely seen in wines not matured in oak barrels. It is lower in acidity thn most Verdicchio, which is good as smoked salmon tends to make wines taste more acidic. It has enough body to balance the fish and it has a superb savoury fruitness that complements it remarkably well.

If you have decided to plump for a more expensive wine, it is probably best to stick to one of the classic areas. You can hardly get more classic than Chablis, in north France. Not only are the wines, at their best, a by-word for elegant refinement, they usually go well with smoked salmon. To do the fish justice, you need a wine that has matured a little and rounded out and one that comes from one of the superior Premier Cru vineyards. Try **Chablis Premier Cru Les Vaillons 1986** bottled by Alain Combard. It has a gently alluring herb butter bouquet and an even-tasting, long-lasting, peachy-appley-buttery flavour. Just as important, though, is that it tastes just as good and just as classy with the salmon.

Along the Loire

Alternatively you could go for Sancerre or, even better, Pouilly Fumé, which comes from just across the Loire river from Sancerre and is similar to it but a tiny bit gentler. Go for **Pouilly Fumé 1988 Les Griottes** produced by Michel Bailly. It has all the gooseberry-like aromas of the Sauvignon grape from which the wine is made, but without the sharpness that accompanies less ripe examples. It is clean, mouthwateringly fresh, with a tang of citrus but also a roundness that gives it subtlety and helps it partner the salmon with aplomb. The classic accompaniment, hollandaise sauce, is easy to make: put 1tsp lemon juice, 1tbls water and seasoning in a bowl set over a pan of hot water. Whisk in 4 large egg yolks, then whisk in 100g/4oz diced butter, piece by piece, allowing each piece to melt before adding the next. Remove from the heat and whisk for 2-3 minutes until creamy.

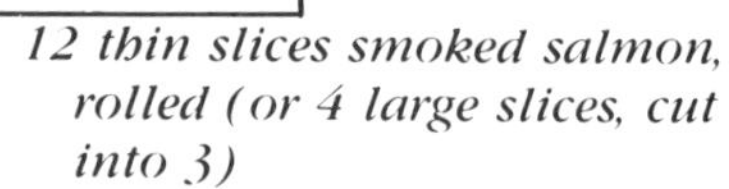

PERFECT PARTNERS

Steamed beans, smoked salmon and hollandaise

This starter illustrates the taste and texture that *nouvelle cuisine* is about. The crunchy beans, steamed until they have just the right amount of 'bite', are a perfect foil for the smokey taste of the salmon. The hollandaise sauce spooned delicately over the beans is light in texture but very tasty.

- ***Preparation: 15 minutes***
- ***Cooking: 30 minutes***

700g/1½lb French beans, topped and tailed
Salt and freshly ground black pepper
175ml/6fl oz hollandaise sauce (see above)
12 thin slices smoked salmon, rolled (or 4 large slices, cut into 3)

- ***Serves 4*** 

1 Place the prepared beans in the top part of a steamer, or in a colander over a saucepan of simmering water. Cover and steam for 15 minutes, until the beans are tender. Season with salt and freshly ground black pepper to taste. Arrange the beans on a heated dish.

2 While the beans are steaming make the hollandaise sauce and keep warm. To serve arrange the steamed beans in the centre of each individual serving dish. Arrange 3 rolls of smoked salmon beside the beans, spoon the hollandaise sauce over the beans and serve immediately.

TAKING THE PLUNGE

KABINETT

Why stick to Liebfraumilch when there are so many other deliciously racy German wines around, all offering a fruity alternative?

FROM ALL THE WINES available to us, German wines are among the best loved. It is easy to see why. They manage to combine lightness of style, freshness and what the Germans call 'raciness' with delicious fruity acidity. Although German wines are usually best on their own, they are also good accompanying food. They are neither too dry nor too sweet for most occasions, nor are they that high in alcohol — making them a pleasure to swallow without being too heady.

It is also easy to stick to Liebfraumilch, probably the best known of all German wines. There is nothing wrong with doing so, but it is a pity as there so many other German wines that you might prefer once you had first taken the plunge and tried them. One of the barriers to trying other German wines is the labelling. The labels tend to be crammed with long, difficult-to-pronounce words, and are often printed in gothic script which makes them particularly hard to read. The way to avoid getting bogged down by all this is to look for the key words, such as Kabinett.

Distinction

Kabinett is derived from the German word for cabinet and originally referred to the wine that was kept as a winemaker's personal reserve: a wine of special distinction. The idea remains, although use of the term has now been formalized in terms of the ripeness of the grapes when picked, as measured by their natural sugar levels. If a wine is entitled to be called Kabinett it will usually say so fairly prominently on the label. One such is **Wiltinger Scharzberg 1983 Riesling Kabinett,** estate bottled by Moselland EG Winzergenossenschaft. It comes from the Mosel-Saar-Ruwer region which covers the lands around the German part of the river Mosel and two of its tributaries. The vineyards grow on steep slopes above the sharply twisting river, creating a breathtaking sight but making cultivation very difficult. Riesling is the grape variety, Germany's most prized, and it gives the wine a delicately floral, slightly honeyed, racy bouquet. The taste is off-dry, mouthwateringly fruity with zingy freshness, leaving the mouth feeling clean and refreshed.

BEST BUYS

Wiltinger Scharzberg 1983 Riesling Kabinett
Taste guide: **medium-dry**
Serve with: **delicate fish dishes**

Steinweiler Kloster Liebfrauenberg Kabinett
Taste guide: **medium-dry**
Serve with: **oily fish, chicken or veal in a delicate sauce**

Morio-Muskat 1986
Taste guide: **medium-dry**
Serve with: **Chinese food, picnic food, sausages**

Liebfraumilch Rheinhessen 1987
Taste guide: **medium-dry**
Serve with: **lightly flavoured dishes**

Richer, grapey and reminiscent of honeysuckle is **Steinweiler Kloster Liebfrauenberg Kabinett,** estate bottled by the Gebiets Winzergenossenschaft Deutsches Weintor. While wines from the Mosel usually come in green bottles, this one has a brown bottle characteristic of wines (like Liebfraumilch) from the various wine regions along the Rhine valley. A little more sweetness on the taste makes it quite full in the mouth but again the crisp, fruit acidity so characteristic of good German wines cuts through and gives an exceptionally lively, pure-flavoured wine — and the flavour lasts and lasts.

THE LANGUAGE OF THE LABEL

The year the grapes were harvested. 1987er means 'of 1987'

The wine comes from a specified region making quality wines. All Liebfraumilch must come into this category

Alcoholic strength

The bottler

Quality control number. All German wines of Q.b.A. category or higher are tested and given an identifying number before being sold

The name of the wine. Indicates wine from certain grape varieties, from specified regions of Germany

The region of Germany from which the wine comes

Country of origin

The contents of the bottle

Variety performance

Until recently German producers have often been modestly reticent about the grape varieties that go to make their wines. Now, though, several are giving this prominence on the label, giving due credit to one of the most important factors in determining the taste of a wine. Buying by grape variety is another way to avoid confusion in selecting German wines. **Morio-Muskat 1986** bottled by Ewald Drathen is an excellent example. Even though not related to the Muscat variety it has much of the Muscat's grapey, musky quality. Delicately aromatic to smell, it tastes lightly perfumed, a touch spicy with clean, green fruit. Though not dissimilar to the better Liebfraumilchs, it does have more character and flavour.

But if you decide that, after all, Liebfraumilch really is the wine for you, at least go for one of the best, such as **Liebfraumilch Rheinhessen 1987** bottled by the enigmatically coded RP342352. It has the softness and creaminess that are the hallmarks of Liebfraumilch but certainly does not lack aroma, fruit or perkiness.

PERFECT PARTNERS

Trout with watercress sauce

- ***Preparation: 25 minutes***
- ***Cooking: 30 minutes, plus chilling***

4 even-sized trout, heads left on
oil, for greasing
squeeze of lemon juice
salt and pepper
orange slices and dill sprigs, to garnish
For the watercress sauce:
1 small bunch of watercress, finely chopped
3tbls mayonnaise
150ml/¼pt soured cream or Greek yogurt
zest of ½ orange or lemon

- ***Serves 4***

1 Heat the oven to 180C/350F/gas 4. Place each trout on oiled foil and sprinkle with lemon juice, salt and pepper. Wrap the foil loosely round each fish and place in a baking tin. Cook in the oven for 30 minutes or until the fish just flakes. Leave the trout in a cool place to cool in the foil.

2 To make the sauce, beat together all the ingredients, seasoning with salt and pepper to taste. Cover and chill until required.

3 When the trout are cold, carefully remove the skin, place the fish on a serving dish and garnish with orange slices and dill sprigs.

4 Spoon a little sauce over each trout and serve the rest in a sauceboat.

KEDGEREE COMPANIONS

CHATEAU DU CHAYNE

The right wine can turn a simple kedgeree into a fabulous feast

KEDGEREE IS A simple, yet remarkably fine dish. Quick and easy to prepare, the combination of flavours leaves nothing to be desired and it is nourishing without being over-filling. This makes it extremely useful for supper or lunch or, its ideal slot, one of the best dishes for brunch. At any of these times, you may decide wine is not called for, but it certainly wouldn't be unsuitable and, if you do plump for a glass to accompany kedgeree, there are a number of suitable wines.

Flavour factors

The more ingredients there are in any dish, the harder it should be to find a suitable wine, as it is quite possible for a wine to go with one of the ingredients yet clash with another one. There are, indeed, several wines which hit this problem with kedgeree. Fortunately, though, despite the fact the wines that you might choose for eggs or butteriness would be different from those chosen for smoked haddock, all these tastes are softened down by the rice and it is not too hard to find wines which do combine well with all the flavours. Kedgeree's flavours are quite delicate and some strongly flavoured wines will overwhelm them, so light or medium-weight wines are the best bet. A good streak of acidity in the wine helps to cut down the richness of the food, but wines that are too steely-fresh do not form an attractive contrast; they appear harsh and aggressive instead.

An example of a lightish white that has just the right sort of weight and acidity is **Baden Dry**. It comes from the Baden region, Germany's southernmost wine-growing district, east of the river Rhine between Heidelberg and Lake Constance. The wines are fatter and fuller in this part than in other parts of Germany although still slight and elegant when compared with the wines of many other countries. Baden Dry has a pleasant aroma of spring flowers and crisp lychees and pears, with a clean, fresh, florally-fruity, dry taste that is not too incisive and lasts in the mouth a long time. The fruity acidity is just enough to keep the wine lively. It cuts cleanly through the kedgeree, without its taste being dulled for a moment by the food, in turn it does not dull down any of the kedgeree's fishy, buttery taste.

Classic combinations

A little more crisp acidity, but nothing too sharp, is the hallmark of **Château du Chayne 1988 Bergerac Sec.** Bergerac lies in south-west France along the Dordogne river,

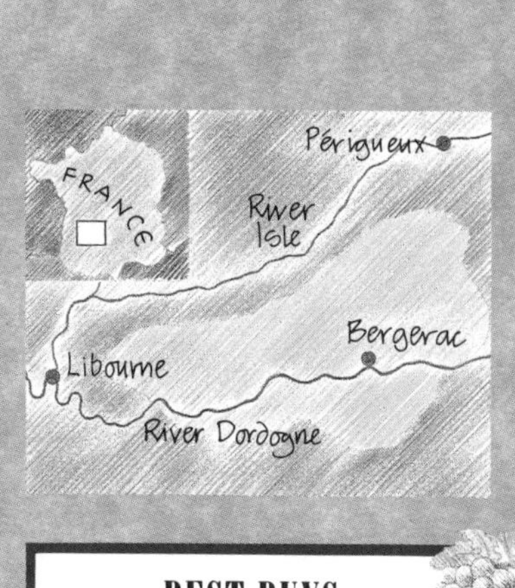

BEST BUYS

Château du Chayne 1988 Bergerac Sec
Taste guide: dry

Baden Dry
Taste guide: dry

La Magdeleine Cépage Terret 1988 Vins de Pays de l'Hérault
Taste guide: dry

San Pedro 1989 Sauvignon Blanc
Taste guide: dry

inland from Bordeaux. It makes wines from similar grape varieties to those used in Bordeaux and often in similar style, though usually with more softness, lightness and elegance. This dry white comes from the Domaine de Bosredon, an estate belonging to the Comte of Bosredon, in family hands and now run by the sixth generation. It is made from the classic white grape mixture of Sauvignon (for acidity and freshness), Sémillon (for fleshiness and roundness) and Muscadelle (for elegance). The predominant impression when it is sniffed is of green fruits: apples, pears, gooseberries and greengages. Then there is a touch of silkiness like lanolin and a hint of brazil nuts. All these flavours come through on the taste in an excellently balanced, mid-weight, lively mouthful, which suits kedgeree and is remarkably refreshing too. Even more important, it brings out and emphasises the kedgeree's flavours.

Just as good a match is **La Magdeleine Cépage Terret 1988 Vin de Pays de l'Hérault.** It is unusual not because of its origin, the far south west of France on the Mediterranean coast, where vines are abundant and great quantities of wine are produced, but because of the grape from which it is made: the white Terret. Although Terret is one of the varieties well at home in France's Mediterranean south, it is not usual for it to be used in its entirety in a wine. La Magdeleine is light, fresh and perfumed, with zingy aromas of fruit peel, herbs, liquorice and acacia. Fresh tasting too, with well-rounded, aromatic fruit flavours, the flavour lasts in the mouth, even against the richness of kedgeree. The cleanliness and balanced acidity make a good partner to kedgeree, and it cuts through the food well without changing the excellent flavour.

A bottle from Chile

Quite a different wine, but none the less a successful match, is **San Pedro 1989 Sauvignon Blanc** from half a world away. It comes from Chile and from the Lontué Valley, one of Chile's best wine-producing areas. The central part of Chile, where all its wines come from, lacks neither warmth nor sun and San Pedro is generously round and full. What makes it work with kedgeree is the grape from which it is made, Sauvignon Blanc. This adds a necessary kick of acidity to what might otherwise be too big and rounded a wine for the food. What is surprising is that this acidity softens a bit when the wine is drunk with kedgeree, the wine becoming softer and creamier, yet the overall effect of food and wine isn't too rich and has an attractive balance. The food itself should have balanced flavours, too, so when choosing the fish for kedgeree avoid bright-yellow dyed smoked haddock and look instead for naturally coloured fish which has a much milder, less salty taste and is a far better choice when you are drinking wine with the kedgeree.

PERFECT PARTNERS

Golden kedgeree

- ***Preparation: 20 minutes***
- ***Cooking: 40 minutes***

450g/1lb smoked haddock
1 bay leaf
50g/2oz butter
200g/7oz long-grain rice, washed and drained
1tsp turmeric
salt and freshly ground black pepper
2tsp lemon juice
3 hard-boiled eggs, coarsely chopped
2 tbls chopped parsley
2 chopped spring onions (optional)

- ***Serves 4***

1 Poach the haddock gently in 600ml/1pt water with the bay leaf for about 15 minutes, until tender. Drain the fish, reserving the water; then skin, bone and flake the fish.

2 Melt half the butter in a medium-sized saucepan and add the rice and turmeric, stir for 1-2 minutes, then add the reserved water, making it up again, as necessary, to 600ml/1pt. Add 2.5ml/½tsp salt and a grinding of pepper. Bring to the boil, cover and simmer over low heat for 20 minutes, until all the liquid has been absorbed and the rice is tender.

3 Using a fork, add the flaked fish, lemon juice, hard-boiled egg and remaining butter to the rice. Check the seasoning, then re-heat gently, stirring to prevent sticking. Serve sprinkled with the chopped parsley and spring onions.

WINE WITH EGGS

LUGANA SAN BENEDETTO

Elegant dishes made with eggs can be complemented with the right choice of wine

EVEN BEFORE EGGS fell from grace as a nutritious staple of our diet and cooking – before the word 'cholesterol' was as common as 'protein' and well before anyone even dreamed of bacteria infestations - there was one group of people who regarded eggs as dangerous territory and best avoided: people who took their wine very seriously. Eggs don't go with wine was their claim and they studiously excluded eggs from their dining tables. The reason why eggs didn't go, whether it was the eggs or the wine which suffered more from the combination, and which part of the egg or style of cooking did most damage, they did not make clear.

Changing habits

It is just as well that wine drinking was not as common then as it is now and that a lot of people have started to drink wine without being aware of this theoretical taboo, for it really is not too difficult to find good wine-egg partnerships. Indeed, if you are planning to use hard boiled eggs in any way in a dish the worst that could happen to them is that their surprisingly delicate flavour could disappear under the onslaught of even a mediumly flavoured wine – if those flavours stay in the mouth. White wines may be unaffected, or may be softened, losing some acidity, which may even be a good thing in one or two cases!

BEST BUYS

Lugana San Benedetto 1988
***Taste guide:* dry**

Vergine Valdichiana 1988 La Calonica
***Taste guide:* very dry**

Sauvignon Blanc 1987 Mendocino County
***Taste guide:* dry**

Mâcon Villages 1987 Montessuy
***Taste guide:* dry**

Château Bonnet 1987 Entre-Deux-Mers
***Taste guide:* very dry**

Eggs and reds

The main problem comes from the yolk of an egg, which is why you have to choose a wine a little more carefully if a dish involves eggs which are lightly cooked. The softer or more dominant the egg yolk is in any dish, the more it is advisable to avoid red wines, for even the roundest will become nastily metallic.

One of the best examples of a white is **Lugana San Benedetto 1988** from Zenato on the southern shores of Lake Garda in north east Italy, made from the Trebbiano grape variety. Trebbiano is Italy's most widely grown white grape variety and it often produces rather uninspiring wines. Near Lugana, though, a special clone has developed which gives much more flavour. This Lugana is crisp and lively, with a refreshingly appley, greengagey, creamy, slightly nutty flavour. It whistles cleanly through the mouth, removing any tacky feeling there may be remaining from the eggs and seems even fruitier and more refreshing after an eggy mouthful than before. But you could also drink it with great pleasure before a meal or with any number of light dishes.

Also made from Trebbiano and also far from uninspiring is **Bianco Vergine Valdichiana 1988 La Calonica.** It comes from central Italy, from south-eastern Tuscany, just north of Lake Bolsena. It is impressive for its subtlety and its gentle aroma of apples, pears and hazelnuts and its restrainedly soft, mixed fruits taste well balanced by enlivening acidity. The wine loses a bit of this acidity with egg dishes while the softness and appley, pear-like and even slightly orange-like fruit remains. There is still enough acidity, though, to keep the wine fresh and help it cut through the richness of the eggs.

Another excellent choice is **Sauvignon Blanc 1987 Mendocino County** from the Buena Vista Winery. From California where grapes always ripen well, it is quite fully flavoured, richly perfumed and has a heavily fruited, powerful, blackcurrants and gooseberries taste. It doesn't squash out the flavour of the eggs and itself is improved by the match. It seems less richly powerful, more fresh and more crisp.

Burgundy would not normally be the best of choices with egg dishes but **Mâcon Villages 1987 Montessuy,** which is lighter in style than many Burgundies and comes from the southern part of the region, Mâcon, makes a remarkably good partner. It is gently buttery, but has a streak of earthiness, characteristic of wines from Mâcon, and a firm steeliness. While it has little effect on the taste of the eggs, it becomes richer and more buttery and its earthiness disappears – so it tastes like a more northerly (and more expensive!) Burgundy.

Entre-Deux-Mers is the land between the two rivers Garonne and Dordogne which meet, become the Gironde and flow into Bordeaux, in south western France. Although some red wine is made, the name Entre-Deux-Mers is reserved exclusively for dry white wine. Until only a few years ago much of it was not very exciting. Now, with a new crop of inspired winemakers the wine has improved by leaps and bounds. One of the leading producers is *André Lurton* with his **Château Bonnet 1987 Entre-Deux-Mers.** It is light and crisp with a little grassiness, a little steely freshness, a touch of lanolin, a touch of gooseberry, a touch of lemon peel. Once more it is a wine which is proved by its partnership with all sorts of egg dishes. The wine seems to broaden, fill out and the flavour becomes even more impressive when tasted. This still remains true, even when served with relatively robust flavours. And since eggs are rarely served on their own and are often combined with other stronger tasting ingredients, such as the black olives and anchovies in the Provençal anchovy omelettes (below), this crisp, dry white wine makes a good choice.

PERFECT PARTNERS

Provençal anchovy omelettes

- ***Preparation: 15 minutes***
- ***Cooking: 20 minutes***

8 medium-sized eggs
salt and pepper
50g/2oz butter
For the Provençal mixture:
50g/2oz canned anchovy fillets in oil, drained
2 garlic cloves, finely chopped
1tbls olive oil
15g/½oz softened butter
1tbls finely chopped parsley
few drops lemon juice
freshly ground black pepper
For the garnish:
50g/2oz canned anchovy fillets in oil, drained
black olives

- ***Serves 4***

1 To make the Provençal anchovy mixture, place the drained anchovy fillets, garlic, olive oil and softened butter in a mortar or bowl and pound or mash to a smooth paste. Add the parsley and season to taste with a few drops of lemon juice and a little fresh ground black pepper.

2 To prepare the garnish, cut each anchovy fillet lengthways into 3 even-sized strips.

3 Break the eggs into a bowl and stir vigorously with a fork or wire whisk to mix the yolks with the whites. Add the anchovy mixture, taste, and season carefully with salt and ground black pepper.

4 Heat a 15cm/6in heavy-based frying-pan over a moderate heat, add a quarter of the egg mixture. Cook the omelette, lifting the edges to permit the liquid egg to run underneath and shaking the pan to prevent sticking. Fry until golden on the underside. Turn the omelette on to a plate and scrape off any bits adhering to the pan. Slide the omelette back into the pan to brown the other side. Keep the omelette warm while you repeat this process to make the remaining 3.

5 Quickly garnish with a lattice-work of anchovy strips and black olives. Serve immediately.

CHEERS TO CHICKEN

RASTEAU

More versatile than most meaty roasts, roast chicken can be beautifully matched with a wide range of fruity reds and whites

CHICKEN IS RATHER an amenable thing to eat. It is suited to a wide range of accompaniments and trimmings, from strongly flavoured stuffings and pungently fruity sauces to delicate creamy sauces or just a simple gravy and plainly cooked vegetables. It is light enough to form the centrepoint of a delicately flavoured dish and assertive enough to play the same role where the other constituents have robust tastes. It should therefore come as no surprise that it can be successfully partnered by many different wines. Indeed, its wine leanings have gone as far as to turn coq au vin into one of its best-known incarnations. Supposedly, the wine to use for this dish is Chambertin, which is extremely expensive, much prized and very rare. But those who stick with such precision to an ingredients list clearly have more money than sense.

Of all the ways of cooking chicken, however, it is roast chicken that is most interesting from a wine-matching point of view. For we tend to think of roasts as meats that take red wines, while chicken is more often reckoned a white wine partner. The fact is that either red or white can make an excellent match. The choice between them could depend on whatever else you might be serving before or after – or just on your preferences.

As good as gravy

One grape variety which does appear particularly well suited to roast chicken is white Chardonnay. The rich butteriness of wines from this grape seems to perform the same function as a rich, succulent sauce or gravy, especially if you have to (or prefer to) eat your chicken without its skin. One of the best balances is provided by **Mâcon Villages 1987** bottled by A Montessuy. It is from Burgundy, towards the southern part of the region and, more precisely, from a group of villages around the district of Lugny, one of the best parts of the Mâcon area. Its buttery nose has a touch of clean earthiness, typical of the area, and it avoids being too fat. It is dry, medium-bodied, with good concentration of flavour, firm and quite rich. But it isn't too overpowering for the lighter tastes of the flesh from the breast, and is incisive enough to cope with the leg meat.

For those who reckon that roast chicken without its skin just isn't worth eating, or who make a beeline

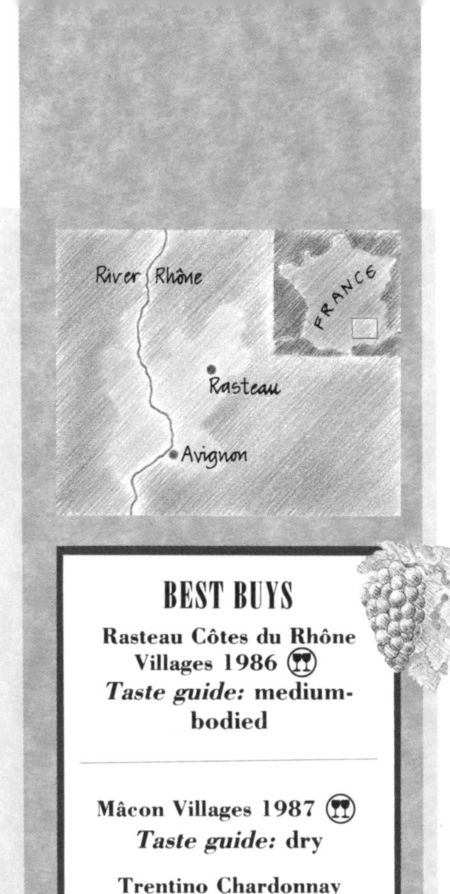

BEST BUYS

Rasteau Côtes du Rhône Villages 1986
***Taste guide:* medium-bodied**

Mâcon Villages 1987
***Taste guide:* dry**

Trentino Chardonnay 1988
***Taste guide:* very dry**

Domaine La Croix Belle 1988, Vin de Pays des Côtes de Thongue
***Taste guide:* very dry**

Señorio de los Llanos Reserva 1981 Valdepeñas
***Taste guide:* light-bodied**

for the parson's nose, a crisper, more acidic wine will make the best match. North east Italy provides the best solution as it produces wines from Chardonnay which share the grape's chicken-friendly characteristics with a light, fresh, fat-cutting style. From the province of Trentino in the far north, not far from the Austrian border, comes **Trentino Chardonnay 1988** from Cantine Mezzacorona. Clean and refreshing, with apple- and melon-like fruit, it will liven up the whole dish.

With or without skin

One of the most exciting roast chicken partners though, is, surprisingly enough, not from Chardonnay but from the white Grenache grape grown in the Herault district of southern France. Called **Domaine La Croix Belle 1988, Vin de Pays des Côtes de Thongue,** it is great whether the chicken is with or without skin, whether the meat is light or dark and whether the sauce or gravy is strong or light. It is remarkably fresh, with a bouquet reminiscent of apples and greengages, but in the mouth has enough roundness and weight to make a creamy balance to the crispness of the fruit flavours. It is this mix of crisp acidity and smooth, rounded silkiness that makes it the ideal partner to all parts of the bird.

For a red to go with roast chicken, it is quite important to choose one with enough fruit. Apart from the fact that fruity reds are attractive, they perform the same function with chicken as fruity sauces. Anything too full-bodied and strongly flavoured runs the risk of overwhelming the poultry, so light and medium bodied wines are certainly the best bets.

The choice is red

A good partner is **Señorio de los Llanos Reserva 1981 Valdepeñas.** Valdepeñas is a large area in central Spain, responsible for a wide array of wines of extremely variable quality. Señorio de los Llanos is one of the most highly respected names – rightfully so – and this well-matured Reserva wine is softly oaky and lightly peppery on the nose, and dry, gently spicy and lightly oaky in the mouth, without too much tannin but with plentiful, gentle, ripe fruit. It is particularly good with the skin and the darker meat.

For a really exciting, fruit-packed wine, there is **Rasteau Côtes du Rhône Villages 1986** from Cave du Grand Jas. The Côtes du Rhône can produce pretty light, easily fruity wines, or big, robust ones. This wine is neither light nor robust but rich, ripe, rounded, full of lively strong plummy fruit, with pepperiness too. Yet despite the richness of its flavour it is not too intense for the poultry and it is a brilliant accompaniment to all sorts of chicken dishes, from flavourful casseroles to pan-fries. However, there is no doubt that this wine is especially good with roast chicken, even better if it has a well-flavoured sauce as in roast chicken with bacon (see below).

PERFECT PARTNERS

Roast chicken with bacon

- ***Preparation: 20 minutes***
- ***Cooking: 2 hours***

oil for greasing
2kg/4½lb chicken
15g/½oz butter
salt and pepper
½tsp dried thyme
8 bacon rashers
225g/8oz mushrooms, sliced
2 large onions, coarsely chopped
2 large cooking apples, cored and sliced
125ml/4fl oz dry white wine
125ml/4fl oz chicken stock
1tsp cornflour blended with 1tbls wine

- ***Serves 4-6***

1 Heat the oven to 200C/400F/gas 6 and grease a small roasting tin. Rub the chicken all over with butter, season generously and sprinkle with thyme. Roast for 10-15 minutes. Lower the temperature to 180C/350F/gas 4, cover the breast with 4 bacon rashers and roast for 45 minutes.

2 Meanwhile, roughly chop the remaining bacon, and mix with the mushrooms, onions and apples. Add the bacon mixture to the roasting tin and return to the oven for 20 minutes.

3 Pour the wine and stock over the chicken and roast, basting occasionally, for 30 minutes or until the juices run clear. Transfer the chicken to a warmed serving dish and arrange the bacon rashers on either side. Keep warm.

4 Drain the juices from the roasting tin into a small saucepan, reserving the bacon mixture. Remove the excess fat. Add any juices from the serving dish, together with the blended cornflour. Cook for 5 minutes, stirring constantly, until the sauce is thick and smooth. Add the bacon mixture and stir well. Serve the chicken hot, with the sauce handed round separately.

OUT FOR A DUCK

LUGANA 1988

Duck with fruit sauce can meet its match with both red and white wines

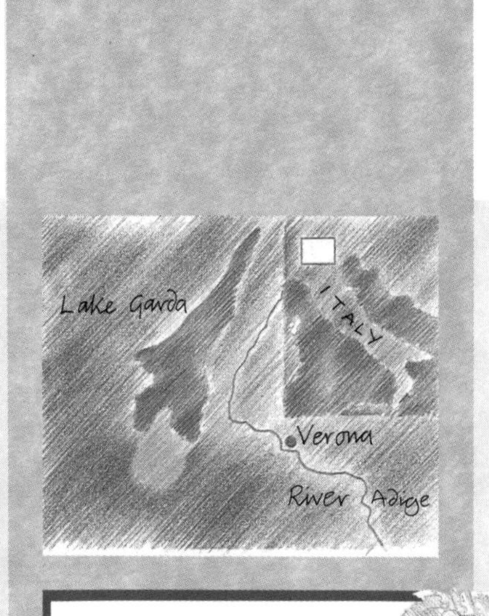

WHILE IT MIGHT be relatively easy to find a wine that is a good match for duck and not too difficult to find wines that go well with fruits, the sort of wines that match duck are unlikely to be the ones you would choose to go with oranges, cherries etc. This means that when serving a duck with one of the fruit sauces that go with it, a little more thought than usual has to go into the choice of wine.

White is right

It is not too difficult, though, to find a wine that matches both the rich, almost gamey flavour of duck and the sweet-sharpness of whatever fruit is chosen to go with it. Indeed, there is even the benefit that a number of white wines which normally wouldn't be the best choice for a simple roasted duck form a splendid combination when the bird is cooked with fruit. The choice is wide, from the widely available Vin de Pays des Côtes du Gascogne for those who go for crisp, dry wines, to Riesling Kabinett wines from Germany, for those who like more fruity sweetness. An ideal match, though, is **Beaujolais Blanc 1987**. Beaujolais, in central eastern France. usually produces red wine, made from the fruit-packed Gamay grape. But there are some white grapes planted in the area too, mainly the prestigious Chardonnay variety, which make superb wines just a few miles further north in the region of Burgundy. Quite deep in colour, it has an encouragingly rich, buttery, appley bouquet and a clean, ripe, quite powerful, rounded taste. It has just the right amount of freshness to remove any sense of fattiness from the duck and just the right amount of richness and weight to balance out the sharpness of the fruit.

Floral freshness

Another terrific match is **Lugana 1988 San Benedetto Zenato**. Lugana is the name of a zone in Italy, on the southern borders of Lake Garda, in the north east, making elegant wines from the Trebbiano di Lugana grape variety. This example is particularly fine as it was made in the excellent 1988 vintage and produced by Zenato, one of the best Lugana estates. It is a delicate wine, light and floral with refreshing acidity adding another dimension to its soft flavours. It is the perfect mouth cleanser to the duck, balancing its rich flavours and complementing the fruitiness of the sauce.

BEST BUYS

Lugana 1988 San Benedetto Zenato
Taste guide: dry

Beaujolais Blanc 1987
Taste guide: dry

Vouvray 1988 Domaine Gauchier
Taste guide: medium-dry

Navarra Cirbonero
Taste guide: light-bodied

Le Canne Bardolino Classico Superiore 1987
Taste guide: medium-bodied

Fruit-sauced duck is also much better than many other dishes with sweeter wines, as long as they are not *too* sweet. One that gets the balance just right is **Vouvray 1988 Domaine Gauchier**. Vouvray comes from the Loire Valley of north-western France and is one of the most famous wines produced from the Chenin Blanc grape, which can make anything from bone dry wines to lusciously sweet ones. Usually, it produces medium-dry wines, with just enough sweetness to take away the sharpness of its strongly acidic character. The wine has a delicate and entrancing bouquet, floral but with a touch of honey and hints of lanolin. It tastes remarkably refreshing with its plentiful acidity, yet rounded, delicate, floral and just a little sweet. With duck, it tastes a touch dryer and gives the fruit in the sauce an extra tang which gives an ideal balance to the gamey richness of the duck.

Red from Navarra

For a red, it is best to go for something fairly light and fruity. Anything too big, full-bodied or tannic clashes with the fruit in the sauce. An ideal choice is **Navarra Cirbonero**. It comes from Navarra, **in the north east of Spain, south of** the Pyrenees and is made by the highly reputed Chivite family. Light, soft and very open, it has cherry and orange flavours that make it a natural for sauces based on these fruits. But it is also gentle and rounded and has the additional flavour of vanilla from its maturation in traditional oak casks. It's a joy to drink and cleverly balances all the varied tastes of the duck whilst its own taste comes through clearly.

For a red with a touch more weight, go for **Le Canne Bardolino Classico Superiore 1987**. Bardolino, from the eastern shores of Lake Garda, is often a very light-bodied wine for drinking young. But this example, from the esteemed house of Boscaini, comes from the Classico heartland area, from a single vineyard and is 'superiore', indicating just a little more alcohol than normal. Although its bouquet is quite meaty, it is nevertheless packed with rounded but lively fruit and finishes dry. It provides a good contrast and becomes a refreshing accompaniment to duck, whether served with a citrus sauce as in duck à l'orange (see below) or with a delicious peachy variation. To make duck with peaches, replace the orange segments with 6 peaches. Peel them, by blanching in boiling water for 1-2 minutes, then halve and stone. Sieve four of the peaches into the sauce at the end of step 4 and simmer for a few minutes. Arrange the remaining peach halves around the duck to warm them through.

PERFECT PARTNERS

Duck à l'orange

- ***Preparation: 25 minutes***
- ***Cooking: 1¼ hours***

1.8kg/4lb duck, dressed weight
salt and freshly ground black pepper
25g/1oz butter
150ml/5fl oz Cointreau
1tbls wine vinegar
juice of 1 orange
150ml/5fl oz beef stock
2tsp cornflour
4 medium-sized oranges, peeled and cut into segments
watercress

- ***Serves 4***

1 Rinse the duck inside and out with warm water and dry with a clean cloth. Sprinkle inside the cavity with salt and pepper, prick the skin thoroughly all over with a skewer.

2 Melt the butter in a deep flameproof casserole, just large enough to contain the duck, and sauté until golden on all sides. Reduce the heat and continue sautéeing gently, covered, for 30 minutes, turning the duck from time to time. Add two-thirds of the Cointreau and allow to simmer for a few more minutes.

3 Remove the duck from the casserole. Skim the fat from the juices then return the duck to the casserole. Add the wine vinegar, orange juice and beef stock to the casserole and bring to the boil. Cover, lower the heat and simmer gently for 20 minutes.

4 Just before serving, remove the duck and keep it hot on a warm serving dish in a low oven. Place the casserole over a high heat and bring the liquid to the boil, stirring constantly, scraping all the crusty bits on the sides of the casserole into the sauce. Reduce the heat and simmer gently for 10 minutes. Skim the fat from the surface and pass the sauce through a fine sieve. Season generously and add the remaining Cointreau.

5 In a small pan blend the cornflour with some of the Cointreau-flavoured sauce until smooth. Pour on the remaining sauce and place the pan over moderate heat. Bring to the boil whisking the sauce constantly, then lower the heat and simmer sauce for 4-5 minutes or until slightly thickened. Add half the orange segments and simmer the sauce a little longer to heat through. Garnish with orange segments and watercress.

GAME CONTENDERS

BURGUNDY

Limited by season and often considered a luxury, game is a rare treat – make the wine rise to the occasion!

HOW MANY TIMES do you see on the back label of some big, full-bodied red, or on a 'shelf talker' or read in a wine article, 'Drink with roast meats and game'? Enough it would seem, to get the impression that there are a huge number of wines that go well with game. Yet how often do you eat game? The game season is restricted (although deep frozen birds are now available all through the year) but it can be expensive to buy.

So there aren't many occasions to try the varying contenders for the role of ideal wine to partner game. It would be a shame to make a mistake when serving game.

Classic choice

It isn't always necessary, or even wise, to go for the deepest, fullest-bodied, most intensely flavoured wine you can locate. Even with the gamiest of game, strong-tasting grouse, there is a delicate, sweet side to the meat apart from its gaminess and a real block-buster of a wine can leave you wondering whether the bird ever did taste of anything or not. One of the classic full-bodied red and game partnerships that does work – particularly if the game is served with all the normal, strongly flavoured trimmings – is with Châteauneuf-du-Pape. It comes from southern France, around the Rhône valley, and the vines are grown in an otherwise infertile soil, composed mainly of large stones. They act as a night storage radiator, storing up the sun's heat by day and reflecting it back by night.

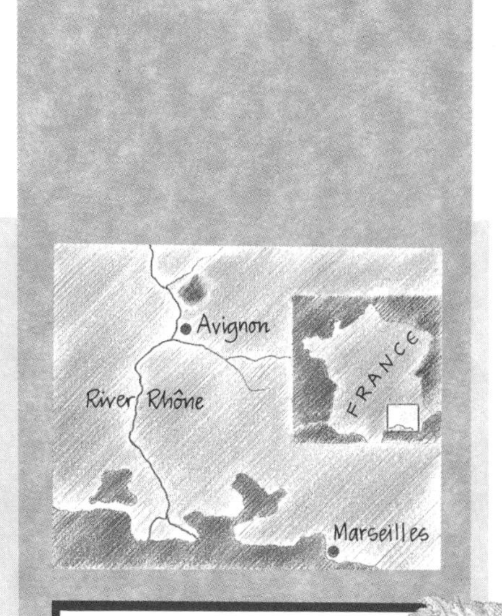

BEST BUYS

Burgundy Pinot Noir 1986
***Taste guide:* medium-bodied**

Châteauneuf-du-Pape 1986 Les Arnevels
***Taste guide:* full-bodied**

Châteauneuf-du-Pape 1986 Les Couversets
***Taste guide:* full-bodied**

Seaview South Australian Cabernet Shiraz 1986
***Taste guide:* full-bodied**

Bourgogne Pinot Noir 1985 Labouré Roi
***Taste guide:* medium-bodied**

Château Tour des Combes 1985 St Emilion Grand Cru
***Taste guide:* medium-bodied**

To see why the partnership with game works, try **Châteauneuf-du-Pape 1986 Les Arnevels.** It is made from a blend of four grape varieties, each of which contributes to the taste of the wine. For example, Grenache gives it strength, body and pepperiness, while the Syrah contributes richness of fruit. The pepperiness acts rather like a condiment to the game, while the strong, pruney fruit rounds out the meat's pronounced flavour. Similar is **Châteaunef-du-Pape 1986 Les Couversets.**

Another full-bodied red that works well with game is **Seaview South Australian Cabernet Shiraz 1986.** There's plenty of fruit, with such ripeness that the wine almost appears sweet. It, too, is from a blend of grapes – just two this time. It gains a blackcurranty nature from the Cabernet and a strong, plummy characteristic from the Shiraz, which also gives it a leathery-fruity smell. The predominanace of the fruit and its sweetness is the clue to its good partnering of game. After all, games goes excellently with sweetish sauces and relishes, particularly when these are fruit-based. The wine is excellent value too.

Burgundy red

It is the sweet character of the fruit, even though the wine may be dry, which is the major factor influencing one of the best, and possibly one of the most surprising, wine-game partnerships of all time: red Burgundy. Red Burgundy, from eastern France, and made from the Pinot Noir grape variety, is not particularly full-bodied or heavy, nor necessarily particularly rich or intensely flavoured. Indeed, Pinot Noir often produces wines which are more delicate than robust, with a soft raspberry character to their flavour. You would have thought that grouse and its cousins would have been far too incisively flavoured for such a wine, yet they are not. In one of those splendid partnerships that defy all logic, the flavour of neither Burgundy nor game is lessened.

Burgundy Pinot Noir 1986 bottled by A Montessuy is a delightful, lightweight, summer fruits (strawberries as well as raspberries) type Burgundy, with a good level of acidity but not too much tannin. Light enough to go with a wide range of starters, with game it takes on a new dimension and epitomizes the brilliance of the match.

Ripe and rounded

A little fuller and fatter is **Bourgogne Pinot Noir 1985 Labouré Roi.** It almost has a slight gamey tang to its rounded, ripe, raspberry fruit. If you have prepared pungently flavoured trimmings, or a rich sauce, the greater richness of this wine would be the better one to use.

You may decide that, given the importance of a game dinner, you want the cachet of a good Bordeaux red to accompany it. If so, then you would be wise to go for something from the Bordeaux district of St Emilion, where the predominant grape is the softer, fleshier, sweeter Merlot (instead of Cabernet Sauvignon which predominates in several other districts). The Merlot's dried fruits and cake character works well with game, as can be seen with a bottle of **Château Tour des Combes 1985 St Emilion Grand Cru.** Grand Cru signifies that the wine comes from an estate producing superior quality and Tour des Combes is indeed classy. But it is not *too* expensive and it will be a great accompaniment to a special meal – whether you go for a tender, plainly roasted game bird, such as pheasant, partridge or wild duck, a rich and warming pot-roast of venison, tasty casseroled rabbit or pheasant, or a simple braised dish, such as pigeons with peas (see below). For more homely game dishes that are eaten on less special occasions (often during the game season), choose one of the less expensive, yet rounded and flavourful reds.

PERFECT PARTNERS

Pigeons with peas

- ***Preparation: 25 minutes***
- ***Cooking: 1¾ hours***

4 young pigeons, oven-ready
salt and pepper
fresh sage leaves (optional)
flour, for dusting
25g/1oz butter
2tbls oil
1 onion, finely chopped
4 thick, unsmoked streaky bacon rashers, cut into strips
150ml/¼pt dry white wine
150ml/¼pt chicken stock, more if needed
1lb shelled peas
boiled small carrots, to serve

- ***Serves 4***

1 Season the pigeons inside and out with salt and pepper. Insert one or two sage leaves into the cavities, if using. Dust the birds with flour.

2 Heat the butter and oil in a heavy frying pan or wide flameproof casserole large enough to hold the birds side by side. When hot, fry them briefly, turning to brown lightly all over. Lift out and reserve.

3 Sauté the onion and strips of bacon in the fat remaining in the pan for 5 minutes. Add the wine and stock, bring to the boil and allow to bubble briskly for a minute. Replace the birds, cover tightly and simmer over very low heat for 1-1¼ hours or until tender. Check the liquid from time to time, and top up with a little extra stock if necessary.

4 Add the peas and simmer for 15 minutes or until the peas are cooked. Take out the birds. With a slotted spoon, transfer the peas, bacon and onion to a hot serving dish. Arrange the pigeons on top. Meanwhile, boil the cooking juices rapidly until well reduced, then spoon over the birds. Arrange the carrots on the dish.

FLOWERY WHITES

GEWÜRZTRAMINER

If a white wine is described as 'flowery' it won't taste like a bunch of chrysanthemums – rather like a taste of summer

IT'S LOVELY TO think of a wine as *flowery* but very hard to define what you mean by that because no wine actually tastes like a flower – as far as we know, because few of us ever eat flowers! Flowery as a description actually hails from the bouquet of a wine, but since much of what we think of as flavour is really smell it is easy to think that the wine actually *tastes* flowery too.

Usually, but not always, the most flowery wines are those made from the grape varieties called 'aromatic' – those with lots of aroma such as Riesling and Sylvaner. These varieties are grown most often in cooler climes such as Germany, England and northern France, especially Alsace. Indeed, there was an advertising campaign for Alsace wines many years ago which showed a large bunch of colourful flowers shooting out of a wine glass. It encapsulated the wine perfectly, as the variety being publicized was Gewürztraminer.

Two characters

The aromatic nature of Gewürztraminer is clear from its Italian translation, Traminer Aromatico. The English translation is more prosaic, calling it Spicy Traminer. Sometimes the wine does seem powerful and spicy, at other times its true floral character predominates, as in the **Vin d'Alsace 1986 Gewürztraminer** from Les Producteurs Réunis de Pfaffenheim. This wine has an attractive, quite deep straw colour, as it should – Gewürztraminer never gives a very pale wine. As you raise the glass to take a sip, the perfume is instantly noticeable. Rather opulent, there's fruit, just like lychees, but also masses of flowers, giving a sense of lying in the garden in high summer. Full and rich tasting, but not sweet, the floweriness could easily get lost under the weight of the wine. But it doesn't – and that's what makes the wine so lovely.

The grape capable of producing the most flowery wines of all, though, is the underrated, versatile Muscat. It may be hard to imagine

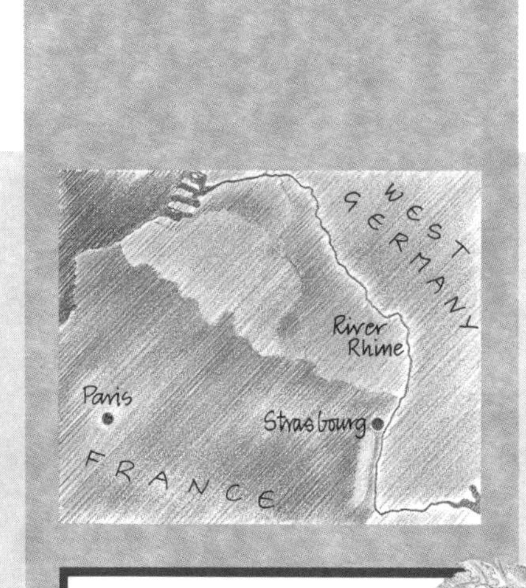

BEST BUYS

Vin d'Alsace 1986 Gewürztraminer
Taste guide: dry
Serve with: fish and meat pâtés and terrines

Moscato d'Asti 1987 Michele Chiarlo
Taste guide: sweet
Serve with: fresh fruit or not-too-sweet patisserie, or on its own

Glen Ellen Proprietor's Reserve White 1987
Taste guide: medium-dry
Serve with: sole, plaice, oily fish; poultry with fruit sauces

Casa La Teja 1987
Taste guide: very dry
Serve with: salad starters, breaded fish or white meat dishes (no cream)

Muscat as a floral variety when drinking the heavy, sweet dessert Muscatels and Moscatels (Muscat's name varies depending on which country it is grown in). To come into its own, the vine has to be grown in Italy (where it's called Moscato) and, more specifically, in the north west. There, not far south of the Alps, the winters are still cold, but the summers quite hot, giving the grape just the conditions it needs to develop masses of perfume. A sip of **Moscato d'Asti 1987 Michele Chiarlo** should be enough to convince. Pale-coloured, with just a light beading of bubbles, the smell is delicate in character but as pervasive as walking through the Chelsea flower show – where the roses are. In the mouth it is light weight, really delicate and on the sweet side, with a flavour of freshly crushed grapes and rose water. Thank goodness its alcoholic strength is low as it begs to be drunk.

Flower power

It isn't just the nature of the grape that can make for a floral wine. Floweriness can turn up from the most surprising quarters if the grapes are harvested at the right time and the winemaking is carefully tuned to harness their aroma to the full. Hot climates are not the natural homeland for floral wines. But you wouldn't realize that from drinking **Glen Ellen Proprietor's Reserve White 1987** from California. Pale and bright, the smell is fresh and clean and gently fruity, but overlaid with flowers; this time delicate spring flowers. Soft and rounded to drink, the floral quality is still noticeable across the peachy flavour, which lasts well after it has been swallowed.

Spanish blooms

Spain is another country not known for flowery whites. Traditionally dark, heavy and flat, in recent years winemaking technology has revolutionized the styles of wines appearing, creating many fresh, squeaky-clean, easy-drinking examples, but not, in most cases, generating floweriness.

The mould-breaker is **Casa Le Teja 1987** from the central Spanish area of Manzanares in La Mancha. The impression on the nose is of wild hedgerow flowers and blackberries, while the taste is dry, quite firm, clean and crisp. Then, just as you swallow, the floweriness comes back. And – at a very affordable price – this wine is good value too.

FOOD FOR THOUGHT

All these flowery wines go well with food that is not overly mild, but not too robustly flavoured either. Lightly smoked chicken, which you will find in good delicatessens and food halls, makes a perfect partner (see below).

PERFECT PARTNERS

Smoked chicken with cherries

- *Preparation: 35 minutes, plus overnight marinating*
- *Cooking: 35 minutes*

450g/1lb fresh black cherries
150ml/¼pt white wine
2tbls oil
1.1kg/2½lb whole smoked chicken, cut into four pieces
salt and pepper
1tsp sugar (optional)
2tsps cornflour
flat-leaved parsley, to garnish (optional)

- *Serves 4* 

1 Remove the stones and stalks from the cherries. Put the cherries in a dish and pour the wine over them. Cover and leave to marinate overnight in the refrigerator.

2 Heat the oil in a large flameproof casserole or heavy-based saucepan and fry the chicken on both sides. Season with salt and pepper to taste. Cover and cook over a very low heat for 20 minutes.

3 When the chicken is thoroughly heated through, add the cherries and wine. Sprinkle with sugar, if using, stirring this in gently with a wooden spoon.

4 Blend the cornflour to a smooth paste with a little water, then add this mixture to the pan. Bring to the boil, then lower the heat and simmer for a further 5 minutes or until the cherries are tender. Remove the chicken pieces from the pan with a slotted spoon.

5 Arrange the chicken pieces on a heated dish and garnish with the cherries. Pour the pan juices over and serve immediately, garnished with flat-leaved parsley, if wished.

FULL-FLAVOURED WHITES

ARROYO BLANC

Not all white wines are pale and interesting; some pack a punch. So for full-flavoured, weighty whites, here's the pick of the bunch

THERE ARE TIMES for pretty little white wines, light and dainty as filigree; there are other times for plenty of flavour. Forget elegance and delicacy, you think, just give me something I can't ignore, something that packs a punch, that fills my mouth with its taste.

There are a number of ways to select wines that will meet such hankerings. You can look for wines made from grape varieties like Chardonnay that usually have plenty of flavour and weight; go for wines from the so-called New World (Australia, California and so on), where the climate is warmer and the grapes carefully cultivated to pack in more taste; choose wines that have spent a bit of time maturing in oak casks – only the sturdiest wines are subjected to this; look for wines with at least two years age; or all four of these characteristics.

Good and affordable

One of the most delicious mouthfillers is **Rully Les Thivaux 1986.** Rully is a village in Burgundy, in the Côte Chalonnaise. It lies just south of Burgundy's prized Côte d'Or and makes wines almost as good, if not as good, as its famed neighbour – but they are more affordable. Les Thivaux is the name of the particular vineyard in the village.

White wines from Burgundy are among the fullest of all France, and the best rank among the biggest in the world. Burgundy is where the Chardonnay grape is most at home and where it produces the rich, voluptuous, buttery tastes that so many other countries try to copy, more or less successfully.

Rully Les Thivaux is a bright, deep straw yellow. It has a wonderful 'drink-me' nose of concentrated, creamy butter, with perhaps a touch of herbs. The same rich butteriness fills your mouth and, to ensure the wine isn't too heavy and leaves the mouth feeling clean, there is just the right amount of refreshing acidity. Try it with rich chicken dishes, fish in cheese sauce, salmon or other 'smart' fish with hollandaise.

Full-flavoured winner

Something quite different, and very special, is **Jekel Vineyard Arroyo Blanc 1985.** Jekel Vineyard is in California, where grapes have no problem in getting good and ripe before picking. This means, as long as the winemaker prunes his vines carefully in winter to avoid rampant over-production and thus dilution, they have plenty of opportunity to

BEST BUYS

Jekel Vineyard Arroyo Blanc 1985
Taste guide: **dry**

Rully Les Thivaux 1986
Taste guide: **dry**

Mainzer Domherr Spätlese 1985 Rheinhessen
Taste guide: **medium-dry**

Brown Brothers Sémillon 1986
Taste guide: **dry**

develop masses of flavour.

The unusual thing about Arroyo Blanc is that it isn't made from Chardonnay, Sauvignon, or any other of the grapes regularly grown in California, but from Pinot Blanc, a grape not always particularly highly regarded in France or Germany. Even in Italy, where it is esteemed, it makes light, steely wines, not huge mouth-fillers. But with the skill of Bill Jekel it's a full-flavoured winner. The colour is quite deep, the nose savoury and spicy, with a hint of vanilla. There's plenty of weight in the mouth – its not inconsiderable 13.5 per cent alcohol adds to the body provided by its flavour, giving 'texture' – and it has a fascinating rounded, creamy, toasty taste with a salty tang that is quite irresistible. Try it with pork.

Best of both worlds

If you love lots of flavour but you like elegance and delicacy too, you can have your cake and eat it. The solution comes from Germany and is **Mainzer Domherr Spätlese 1985 Rheinhessen.** It has all the pretty floral, aromatic fruitiness that is Germany's hallmark, but it is Spätlese, which means that it comes from later-picked, riper grapes, which have developed more flavour.

It won't seem quite as mouth-filling as other, non-German wines, because it has less alcohol, only 8.5 per cent, and alcohol gives a feeling of weight and body to wines. But there's plenty of taste; rounded, fruity, mouthwateringly refreshing, medium-dry taste. Enjoy it to the full on its own, or see how well it goes with many of your favourite meals.

Better and better

Often the concentration of flavour in wines develops only after they have spent some time in bottle, by which time all the wine has usually been sold, and probably drunk. If you'd like to witness how much wine can round out, and if you have a quiet nook or cranny at home, are prepared to invest in the experiment and won't be tempted to raid the spoils, buy a couple of bottles of **Brown Brothers Sémillon 1986** from Victoria in Australia.

The Sémillon grape usually makes sweet wines, like Sauternes, in France, but it is ideally suited to Australia, where it makes full, but dry, wines that can be absolutely glorious. They are often very attractive when young then, a year or so later, seem nice but unexceptional. However, after another couple of years or so they turn into something far, far better. Brown Brothers Sémillon has plenty of fleshy, oily, floral, almost eucalyptus character now, that won't have you complaining for lack of flavour, but wait a while and it will get far better.

The choice of main course to go with these full-flavoured wines is wide. The dish needs to be flavourful, yet not too powerful, and the loin of pork with orange, below, is perfect.

PERFECT PARTNERS

Loin of pork with orange

This joint, which is equally good served hot or cold, has a crisp, spicy crust

- ***Preparation: 30 minutes***
- ***Cooking: 2 hours 10 minutes***

1.4kg/3lb loin of pork, weighed after skinning and boning (keep skin and bones, if possible)
salt and pepper
1-2 garlic cloves, finely chopped
2tbls chopped parsley
1tsp chopped fresh marjoram or ½tsp dried
2tbls oil
200ml/7fl oz hot chicken stock
juice of 1 large orange
3-4tbls orange liqueur
1tbls French mustard
3tbls dry white breadcrumbs
1tbls soft light brown sugar
orange segments and watercress, to garnish

- ***Serves 8***

1 Heat the oven to 170C/325F/gas 3. Sprinkle the inside of the loin with salt, pepper, garlic, parsley and marjoram. Form it into a neat roll, fat outwards, and tie at intervals with string.

2 Heat the oil in a flameproof casserole into which the meat fits closely. When it sizzles, brown the meat lightly on all sides. Add the chicken stock. Tuck the bones and pieces of skin, if available, around, cover and roast for 1¾ hours.

3 Discard the bones and skin. Pour the orange juice and liqueur over the meat, spread the meat thinly with mustard and sprinkle with the breadcrumbs and sugar. Return to the oven and cook, uncovered, for 15 minutes or until the juices run clear and the topping is crisp.

4 Remove the meat from the casserole, cut the string and slice the pork neatly. Arrange on a hot serving dish and keep warm. Degrease the pan juices, add the orange segments and heat through. Garnish the pork; serve with the juices.

WHITE WINES FOR CARNIVORES

CHARDONNAY SANTI

If you like eating meat but you don't like red wine, does this mean you can't drink wine with your favourite dishes? Not at all!

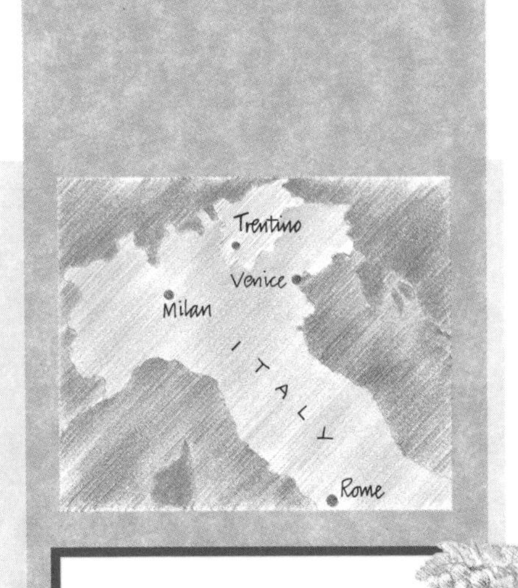

BEST BUYS

Chardonnay Santi 1987
Trentino
***Taste guide:* dry**

Alsace Riesling 1985
***Taste guide:* dry**

Rioja Navajas
***Taste guide:* dry**

Barossa Valley Sémillon 1986 Adelaide Wine Estates
***Taste guide:* dry**

THE MUCH-QUOTED 'white wine with fish, red wine with meat' aphorism may be a rough guide for what wine to drink when, but it's too simplistic and overly restrictive. Basically, you can drink whatever you like with any dish as long as you personally like the combination; there are no hard and fast rules.

It is true, though, that good red wines and tasty hunks of meat form a stunning match. With white wine and meat dishes you will rarely get this 'one plus one equals three' combination; what you will get is an additional complementary flavour, equivalent to the effect of a tasty vegetable, or even good bread.

White winners

Because many meat dishes are pretty robust, it is natural to imagine that a big, full-bodied white is the best partner. Hearty whites do go well, but then so do lighter ones. In fact, it is surprising just how many different styles of white wine do

LEAD CAPSULES

Capsules for wine, often so innocent looking, can be surprisingly difficult to prize from the bottle. The best types hug the bottle tightly and are quite firm and thick, yet flexible and easy to cut through and remove. This is because they are lined with lead.

It is now thought that it is just possible that imperfections on the inside of the capsule can let the lead come into contact with the bottle neck and form lead oxide, which can be washed into a glass as the wine is poured. As no one wants to take the risk of increasing the amount of lead in their body, however minimally, it makes sense to wipe the neck of the bottle with a damp cloth after opening it. This has the added advantage of removing any odd bits of cork dust.

To be even more meticulous, you could ensure you always cut the capsule below the bulge of the bottle neck, or whip it off completely, ignoring aesthetics, to avoid any possibility of wine flowing over the cut, lead-exposed, edge. But don't worry, no wine taster has yet died of lead poisoning!

shine with meat.

For example there's **Chardonnay Santi 1987 Trentino.** Trentino is in the far north east of Italy, along the Adige river valley, surrounded by the soaring Dolomites. In this region the much-loved Chardonnay grape variety thrives, but produces fairly light-bodied wines, with a delicate, lemony, creamy nose.

Tasting Santi's Chardonnay on its own, with its crisp, herby, slightly perfumed, silky character, it is hard to imagine that it could stand up to a meaty main course. Yet even pitched against the strongest, most garlicky salami its flavour remains intact (as does the salami's) and the fresh acidity of the wine complements brilliantly the richness of the sausage. It would be just as good with lamb, and also provide a good match for lighter meat like veal.

Of the world's major grape varieties, Riesling is one of the most reticent to display its glorious character, often needing several years to develop. Unlike in Germany, where the wines from this grape are usually medium-dry or sweet, in Alsace, across the Rhine river and across the border into France, they are dry. **Alsace Riesling 1985** from the Cave Vinicole de Bennwihr is a prime example. It has the luscious, floral character of the grape which, when sniffed, can seem almost like petrol! It has good weight in the mouth, and lots of firm, rich, flowery, but not over-ripe fruit. It is the firmness of this wine which makes it a good partner for meat, especially for something as flavoursome and fruit-attracting as duck. It is assertive enough to cut straight through the food flavours, while any slight hardness will be softened out by the combination of meat and wine.

White velvet

By drinking white in what is thought of as a red wine's domain, you might as well carry the theme to its logical conclusion and drink a white wine which is normally red. Rioja, from north east Spain, is best known for its velvety, oaky reds. White is also made in the area and sometimes it can be just as oaky as its red partner. It can also be very light, fresh and non-oaky if its maker so desires.

A carnivore's best bet is a half-way house like **Rioja Navajas,** which has plenty of lively, sappy fruit but also a strong vanilla-like tang from the use of oak. This is the sort of wine to accompany really succulent, mouth-watering steaks. Just as with red wines, when red meat softens their tannin, so it softens wood flavours in whites, leaving texture and fruit.

White silk

For a fuller-bodied white, but not so full-bodied that it might fight with the food for attention, Australia produces just the right style; despite the claims of the widely distributed Chardonnay grape, the Sémillon is ideal. Mid-weight and rich, smelling of tropical fruit and lanolin, filling the mouth with silky, ripe, full but elegant fruit, **Barossa Valley Sémillon 1986 Adelaide Wine Estates** would be a dream with pork. The label says, 'To enjoy with richer fish or poultry dishes', but that's just following convention - the pork casserole, below, mildly scented with sage, would make an ideal partner.

PERFECT PARTNERS

Pork casserole

- ***Preparation: 20 minutes***
- ***Cooking: 2 hours***

25g/1oz flour
salt and pepper
½tsp dried sage
700g/1½lb lean pork, trimmed and cut into 4cm/1½in cubes
3tbls oil
2 onions, thinly sliced
225g/8oz Jerusalem artichokes, finely diced
2 cooking apples, cored and sliced into 5mm/¼in thick rings
4-5 fresh sage leaves (optional)
150ml/¼pt white wine
For the topping:
450g/1lb potatoes
25g/1oz butter
3tbls milk

- ***Serves 4*** (££)

1 Season the flour with salt, pepper and the dried sage. Toss the pork in it until well coated. Heat the oven to 180C/350F/gas 4.

2 Heat the oil in a flameproof casserole and fry the meat over medium heat until it is browned on all sides. Remove with a slotted spoon and fry the onions for 3 minutes, stirring often.

3 Return the meat to the casserole, add the artichokes, apples and sage leaves, if using; pour on the wine. Bring to the boil, cover the casserole and cook in the oven for 1½ hours or until tender.

4 Meanwhile, cook the potatoes in boiling, salted water for 20 minutes or until tender. Drain and mash them with the butter and milk. Season well.

5 Stir the casserole and skim off any fat. Spread the potato over the top and fork up into peaks. Return to the oven for 20 minutes to brown. Serve hot.

FRUITY WINES FOR BACON

VIN DE PAYS DE L' ARDECHE

The wine you usually choose to accompany a dish may not be right if bacon has been added

YOU WOULDN'T EXPECT wine with your rashers, even on a Sunday morning, nor when you tuck into a bacon sandwich. So why worry about which wines go well with bacon?

Because bacon is an ingredient in many dishes and can make all the difference to how they taste. It can be used to cover the breasts of poultry or game to stop them drying out; added to casseroles or terrines; used as a garnish in all sorts of ways because it adds its own unique flavour to a dish. So it could be that the next time you make a chicken dish you will serve a wine that you've enjoyed before, only to find that it doesn't seem quite right, not realising that the bacon has altered the match.

Bacon is salty, and just as fairly astringent wines cut the richness in foods the bacon's saltiness will cut the richness of wines. On the other hand there's quite a lot of fat in bacon – it contains much of the flavour – so the wine needs to have decent acidity to balance that. To end up tasting the wine's acidity but not its rich fruitiness can be a pretty miserable experience.

Evocative bouquet

One wine that shines through the bacon barrier is **Vin de Pays de l'Ardèche 1987**, bottled by Les Vignerons d'Ardèche. Though it is from southern France, near the Rhône valley, where the summers are hot and the wines can be big and strapping, this one is from a lighter mould. Its bouquet evokes the region itself: warm stones, hot earth, herbs, dry vegetation. But in the mouth it is fruit first – as all wines should be – with the dry, warm, lightly tannic, herby, peppery, tea-like tastes following. Good as it is on its own – or with pâtés, poultry, nut roasts or pork – it tastes even better when there is some bacon in the dish. The bacon enlivens the fruit and makes the wine irresistible.

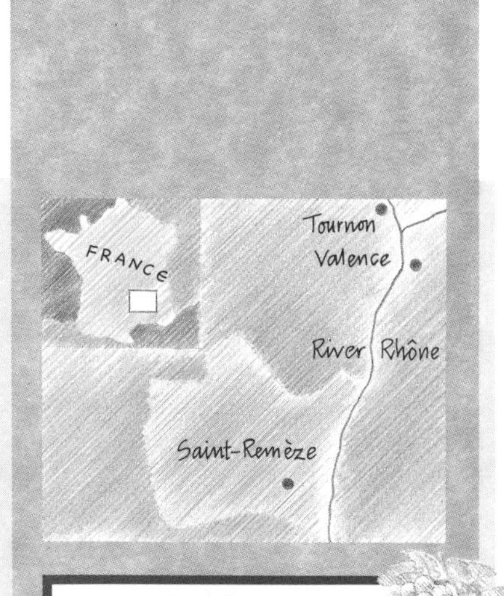

BEST BUYS

Vin de pays de l'Ardèche 1987
Taste guide: light-bodied

Collio Merlot 1980 Collavini
Taste guide: medium-bodied

Cahors 1985 de Domaine de Leret-Monpezat
Taste guide: full-bodied

Australian Shiraz Cabernet Murray Valley
Taste guide: medium-bodied

Kreuznacher Kronenberg Riesling
Taste guide: medium dry

A 'summer pudding' wine

Another splendid bacon matcher is **Collio Merlot 1980 Collavini**, from Collio, on a ridge of hills that lies partly in the extreme north east of Italy and partly in Yugoslavia. The Merlot grape, usually thought of as French, was brought to this part of

Italy many years ago and flourished. The wine, which has aged slowly and magnificently, still has a deep ruby colour. Nor does it lack fruit. Its exotic mixture of blackberries, redcurrants, blackcurrants and raspberries makes it a summer pudding of a wine. There is not too much tannin but lively fruit-acidity, a touch of spice and a soft, rounded feel in the mouth. The salty flavours of bacon cut clean through the wine, tasting even more succulent, while the wine tastes just as impressive but a little more spicy.

Strange but sniffable

Cahors, from the *département* of Lot in south western France, used to be called 'the black wine'. It was so inky deep it was opaque and it took decades to become ready to drink. Now it is made differently and is softer, often much lighter in colour and ready for drinking in a few years. But some examples, like **Cahors 1985 Domaine de Leret-Monpezat**, are still very dark. This wine has the strangest but most attractively sniffable sweet-sour bouquet, rather like a mixture of brambles and greengages. It is firm and full of flavour, with some tannin and balanced acidity, and its dominant fruit is augmented by liquorice, a touch of lemon and a hint of the darkest black chocolate. It makes a perfect foil for bacon, especially when it is part of fully flavoured casseroles or other game or red meat dishes.

Mangoes, leather and oak

Australia too has wines that can take on bacon. **Australian Shiraz Cabernet Murray Valley,** from south Australia, is made with a mixture of Shiraz and Cabernet grapes. It has an invitingly warm and mellow smell of ripe blackcurrants with mangoes and leather, and it tastes soft and velvety, with plummy fruit and a light spicy tang. Its mellowness comes from its having been matured in oak barrels, and it is allowed to take on just enough oak flavour to soften it without overriding its other flavours. The contrasting flavours of each, counter-points the other to the advantage of both. Shiraz Cabernet is particularly recommended with roasts.

A touch of musky spice

What about a white? After all, bacon can be used in white wine-type dishes or you may simply prefer white wine to red. Germany is a good place to look, especially the Nahe region, and **1987 Kreuznacher Kronenberg Riesling** is just the thing. Its fruity acidity ideally complements bacon's salt and it has enough weight to cope with the bacon's strong flavours. It has plenty of soft flowery fruit and a little musky spice makes it good to drink in its own right. With terrines, white meats, salads and many other dishes, it is excellent.

If you are looking for a wine to go with a bacon-based appetiser, such as bacon and liver rolls, below, then this Riesling is just right. The strong flavoured chicken liver filling might overpower other lighter wines, but the Riesling's spicy, fruity flavour makes it very quaffable, both as a pre-dinner tipple and to drink with the meal.

PERFECT PARTNERS

Bacon and liver rolls

- ***Preparation: 20 minutes***
- ***Cooking:15 minutes***

225g/8oz chicken livers
2 hard-boiled eggs, shelled
25g/1oz butter, softened
2tbls finely chopped onion
2tsp finely chopped parsley
salt and pepper
1tsp brandy
½tsp lemon juice
10 streaky bacon rashers
watercress, to garnish
1 tomato, cut in wedges, to garnish

- ***Makes 20***

1 Put the chicken livers in saucepan and cover with water. Bring to the boil, then simmer gently for 5 minutes. Drain.

2 Mince the hard-boiled eggs and chicken livers in a food processor or rub through a vegetable mill. Add all the remaining ingredients, except the bacon, and stir to blend. Cover and chill.

3 When ready to serve, heat the grill to medium. Lay each bacon rasher in turn on a board and run the blade of a knife over it to stretch it. Cut each slice in half and spread thickly with the liver mixture. Roll up and secure with a wooden cocktail stick.

4 Grill for 10 minutes until the bacon is crisp, turning occasionally. Pile the rolls onto a serving dish, garnish with the watercress and tomato wedges and serve.

WINES TO COMPLEMENT SALAMI

COTES DE LUBERON ROSE

A varied selection of wines to accompany both strong and mild flavoured salamis

BEST BUYS

Côtes du Lubéron Rosé Domaine de la Panisse 1988
Taste guide: **light-bodied**

Cavicchioli Lambrusco Amabile
Taste guide: **light-bodied**

Monteleone Soave Classico 1987 Vigneti di Costeggiola
Taste guide: **dry**

Vigneti di Marano 1986 Valpolicella Classico Superiore
Taste guide: **medium-bodied**

Lindemans South Australian Shiraz 1985 Bin 50
Taste guide: **full-bodied**

THERE ARE MANY occasions on which you may find yourself eating salami. It could be on a picnic or as part of a simple lunch – perhaps just bread, cheese and salami; it could be as part of a buffet of cold meats; it might be a constituent of mixed hors d'oeuvres; or perhaps for something to nibble with an aperitif. Whichever role it takes, you will need a wine that matches it. Salami, though, can be pretty powerfully flavoured, or it can be surprisingly delicate, so the ideal is to find a wine of the right weight for a particular salami. If that is impossible – as it may well be without testing both wine and salami first – you need a wine that adapts well to as many different types as possible. No wine will match the lot, but then no wine would adapt to all the varied types of occasions on which salami might be eaten either.

Sparkling Lambrusco

One of the best, and probably the most surprising, salami partners is Lambrusco. It certainly goes with as broad a range of salami as you could imagine. But then Lambrusco comes from Emilia in north central Italy which is the home of a lot of good salami – as well as Parma ham and Parmesan cheese. Lambrusco, whether dry, as it was traditionally, or sweetish, as it more commonly is now, has a good level of acidity, which its sweetness often obscures. That acidity is put to excellent use, though, with salami. It cuts right through the richness and acts as the perfect foil to any over-assertive flavours. Yet Lambrusco itself is not all that fully flavoured so will partner the milder types of salami just as well. It is not upset by garlicky salami nor does it ever taste metallic with them, as some wines do. What more could you ask? **Cavicchioli Lambrusco Amabile** is an excellent example. Deeply coloured, with just a little fizz, it has a gently sweet, meaty, cherry smell with a lively, fresh, clean, firm, sweetish taste. It is low in alcohol (7%) too. It might not be the most suitable wine for a smart platter of cold meats, but it is ideal for a picnic, for a simple bread-and-salami lunch and would be great with salami nibbles on any informal party or other occasion.

Also for the more casual occasions

on which salami plays a part there is **Côtes du Lubéron Rosé Domaine de la Panisse 1988**, from Provence in south eastern France. Although a rosé, it is quite a deep colour, more like a pale red, and so is fine whether well chilled or only slightly cool. It has a firm, rich, strawberry-earthy nose and a dry but fruit-packed strawberries and raspberries taste, with very little mouth-drying tannin. At its best with medium-weight salami, it goes very well too with the coarser, more powerfully flavoured types. The wine would be the perfect choice for mixed hors d'oeuvres where salami played a part, especially since its half-red, half-white character means it would go with a wide variety of meats, vegetables and other foods. It also has enough body to accompany a meat-based main course. And as a partner to a light salami-based meal, whether al fresco or not, it could hardly be bettered.

Aperitif

For an aperitif many people prefer a white wine. It is not always easy to find a white wine which is light enough to drink on its own before a meal and yet will stand up to the powerfully flavoured nibbles you may be serving. For if salami is offered, olives and other incisive tasting bits may be too. The clue is to remember that white wines are usually light and delicate on their own but are made to go with food - and Italian food is rarely lacking in flavour. **Monteleone Soave Classico 1987 Vignetti di Costeggiola** is a great choice. Made by the excellent Boscaini company with great care and attention, it is streets ahead of much Soave. It smells almondy, with creaminess, grassiness and applelyness too. It has a real zing of refreshing acidity - which is just right to cut through the salami as well as enlivening dulled taste buds and stimulating the appetite - and a rounded, lightly nutty taste.

Antipasto

The same producer also makes **Vignetti di Marano 1986 Valpolicella Classico Superiore**. The vineyards of Valpolicella are situated right next to those of Soave in north-east Italy. The same skills are employed and result in a Valpolicella that is also well above the norm. It is a bright but lightish ruby red, with a meaty, bitter cherry bouquet and a clean, fresh, cherry kernel taste with a strongly bitter finish that disappears with food. Its quality, lowish weight and food-needing style makes it just right to accompany meat-containing first courses. As it also goes well with salami, its role in a mixed *antipasto* is second to none.

For those cold cuts which may include anything from chicken to beef, with tongue, mortadella and salami included, a wine has to be pretty versatile and have plenty of its own fruit to complement the meat. **Lindemans South Australian Shiraz 1985 Bin 50** fits the bill excellently. It is deeply coloured, with a rich, meaty, leathery, smoky smell, enlivened by a touch of lemon. Full of flavour, the nature of the fruit is sweet enough not to turn metallic against salami.

When salami is combined with fruit and cheese as in the grilled apple and salami croustades, below, the wine needs to be well flavoured and fruity, yet with enough acidity to cut through the richness of both salami and cheese. Soave is a good choice, if you prefer white wine, or go for the Shiraz, if you like to drink red.

PERFECT PARTNERS

Grilled apple and salami croustades

These small, elegant, apple croustades are ideal for serving with drinks, or even as a savoury at the end of a meal

- ***Preparation: 10 minutes***
- ***Cooking: 10 minutes***

2 small dessert apples
50g/2oz butter
6 large slices white bread
6 slices of salami with the skin removed
6 slices of Cheddar cheese, cut in rounds
paprika

- ***Makes 6***

1 Peel and core the apples with an apple corer or sharp pointed knife. Slice each apple into 3 even-sized rings about 1cm/½in thick.

2 Melt 25g/1oz butter in a frying pan, then add the apple rings. Brown the rings on both sides. Remove them with a slotted spoon and drain thoroughly.

3 Using a pastry cutter or glass, stamp out a circle 6-7.5cm/2½-3in in diameter from each slice of bread. Heat the remaining butter in the frying-pan and when sizzling add the bread circles. Brown on both sides. Remove the croûtons with a slotted spoon and drain on absorbent paper. Heat the grill.

4 Place an apple ring on each fried croûton of bread. Cover each with a slice of salami and top with a round of Cheddar cheese. Sprinkle each apple croustade with a small pinch of paprika and grill until golden brown and bubbling.

ALTERNATIVE PACKAGING

LAMBRUSCO

There are many different ways of buying wine other than in bottles – try wine boxes, cans or cartons for picnic-to-party convenience

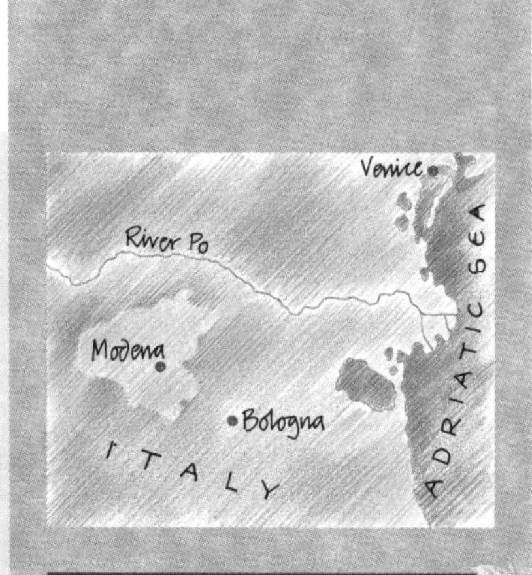

BEST BUYS

Lambrusco Amabile dell' Emilia
Taste guide: **light-bodied**

Deutscher Tafelwein Rhein
Taste guide: **medium**

Pfälzer Landwein (1L)
Taste guide: **medium-dry**

McWilliam's 'Hillside' Shiraz Cabernet (2L)
Taste guide: **full-bodied**

Coteaux d'Aix en Provence (3L)
Taste guide: **medium-bodied**

IT ALL STARTED around the end of the 1970s. A few bag-in-box wines, imported from Australia, appeared – and then disappeared again. Few people in those days seemed prepared to give such unconventionally packaged wine the benefit of the doubt. After a few years bag-in-box wines re-emerged, found tremendous success with the more sophisticated consumers of the 1980s and went on to spawn other forms of alternative packaging for wines, such as cans, cartons and plastic bottles.

The plastic bottle is now seen only occasionally, but boxes, cartons and cans have consolidated their position as legitimate alternatives for the wine buyer. But why buy wine in anything other than bottles? Basically, in certain situations, other containers are much more convenient.

Wine cans hold 25cl, which is less than the standard 33cl fizzy drinks can. 25cl is a third of a bottle, or enough for two decent-sized glasses. This is one of their major benefits: there's just enough without any wastage. Other advantages are that they are light and easy to carry, easy to open, difficult to break, and the wine inside can be chilled (or warmed up) quickly. They are ideal for picnic lunches, drinks outdoors or away from home, or anywhere where ease of accessibility is important, or a corkscrew is lacking. Their major disadvantage is that there isn't a very wide choice available, so you may be unable to find exactly what you want. It is also worth remembering that the technology of canning an unstable product like wine has to be paid for: wine in cans is usually more expensive than the equivalent quantity in bottle.

The pick of the canned whites is a

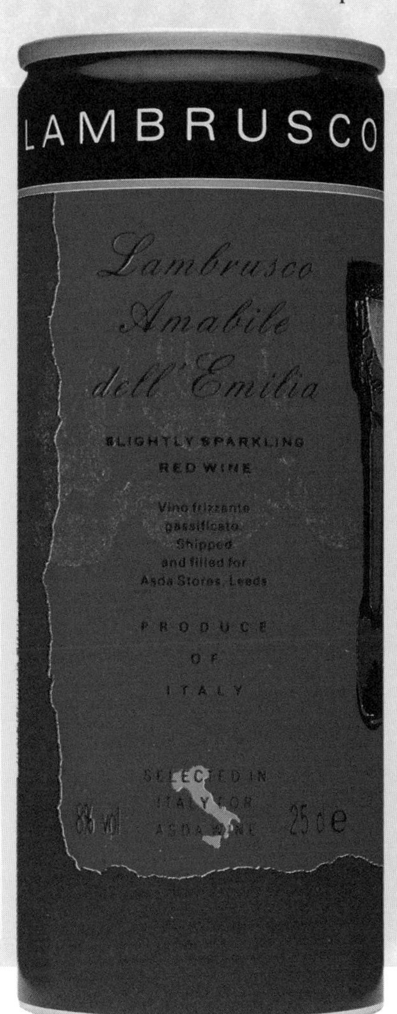

STORING WINE

If you buy wine in a box, carton or can, you won't be thinking about storing it, just drinking it. But if you want to store your bottles of wine, you don't have to unravel yards of complicated mystique about cool, damp, even-temperatured cellars and the like; as long as it's not stored by a radiator, fire or cooker, or in the freezer, the wine won't suffer.

If you are going to store wine at home you should find a reasonably even-temperatured spot and lie the bottles flat (or as near so as you can manage) so the wine stays in contact with the cork. This keeps the cork moist and ensures it doesn't dry out and shrink, which would let in air and spoil the wine. But cork won't dry out all that quickly, so if it's only for a couple of weeks or so it is perfectly all right to keep your wine upright.

semi-sparkling white wine from Germany, a **Deutscher Tafelwein Rhein** which is a basic wine from vineyards around the Rhine valley. It has the attractive, light, flowery fruitiness of most German wines, but the sparkle – nowhere near as much as in a fully sparkling wine – gives it a lift and adds freshness. From the reds, best is a summer wine *par excellence*: Italy's Lambrusco. **Lambrusco Amabile dell' Emilia** is light coloured, a little sweet, a little fizzy, and packed full of the cherry-like fruit of the Lambrusco grape. It is particularly good with salami.

Open here

Cartons, otherwise variously known as Tetra-paks, Tetra-briks or Combiblocs, are the same brick-shaped containers that long-life fruit juice comes in. Just like the fruit juice cartons, they are devilish to open if you haven't got a pair of scissors or a knife handy, and tricky to pour until you get the knack. They are also just as light and easy to carry and practically indestructible before opening. There are some 25cl wine cartons around, which have much the same conveniences as cans. There are also, though, litre cartons. These are only good if there are enough of you around to ensure you finish the wine, for once opened they are not easy to re-close. But for anywhere outdoors, particularly if there are children around, cartons are the safest containers you can find: there's absolutely no risk of cuts from broken glass or sharp can edges. Try **Pfälzer Landwein** which is also Deutscher Tafelwein but Halbtrocken (half-dry). Light, fruity and refreshing, it's German wine at its pleasurable simplest.

Wine in a box

The essence of bag-in-box packaging is the collapsible bag inside the box, allowing you to take as much or as little wine as you want, whenever you want, without it going off. Fine for parties, and certainly easier to lug home than four bottles (the equivalent of the most common 3-litre size), bag-in-boxes are great if you like to stick to the same wine most of the time or won't be tempted to take another glass, and another, every time you go to the fridge or shelf where the box is stored. However, to find out whether you like a certain wine can be an expensive experiment (just about the cost of four bottles) and, whatever the producers claim, the multi-layered foil and plastic bag is just not as good as glass (or can or brik) at keeping out the air – wine's natural enemy. The wine *will* stay roughly the same once open for the four months most manufacturers recommend – but it might not be in peak condition when you open it. So it's important to buy from a supplier with a swift turnover of goods which usually means a major supermarket chain.

Mc William's 'Hillside' Shiraz Cabernet is one of the best wines-in-a-box around. It's a big, rich wine with plenty of spicy, minty flavours. It is eminently drinkable, though, and unusually, comes in a slightly smaller box size - 2-litres instead of the more readily available 3-litre boxes. If a full-bodied wine is a little too much for everyday drinking, go for **Coteaux d'Aix en Provence** - full of ripe, plummy, fruit flavours, and with just a hint of summery herbs, it's great with all meat dishes - try it with the salami and fennel antipasto, below.

PERFECT PARTNERS

Salami and fennel antipasto

- ***Preparation: 15 minutes, plus chilling***

200g/7oz full-fat soft cheese
½tsp coarse salt
100g/4oz Italian or French garlic salami, very thinly sliced, rinds removed
450g/1lb bulb fennel, thinly sliced
pepper

- ***Serves 4*** (!)(££)

1 Mix the soft cheese thoroughly with the salt. Shape the cheese into a roll and refrigerate for 1 hour.

2 Divide the salami between four plates, overlapping the slices around the edge of half the plate. Arrange the fennel slices around the other half of the plate.

3 Cut the cheese roll into eight and put two slices in the middle of each plate. Season generously with pepper and serve.

A SOUPCON OF SPRING

FRANCIACORTA ROSSO 1987

The first lamb of the season will be well enhanced if you choose a red wine with a strong character

THERE ARE FEW things as delicious as a joint of new spring lamb, perfectly cooked, with a hint of pink in the middle and full of succulent flavour. It is a dish that almost cries out for a good bottle of wine to accompany it, for as good as lamb can be it will be even better when washed down with a good wine. A splash of wine in the gravy will also enhance the meal.

Spoilt for choice

It is not all that difficult to find wines that partner lamb well enough, particularly if they are red, but there is one which is an ideal match: **Franciacorta Rosso 1987** produced by Contessa Camilla Maggi. It comes from a small zone of gently rolling hills in central northern Italy, near Brescia, not far from Milan. It is unusual in that it is made from a unique blend of four grape varieties: Nebbiolo and Barbera, which come from Piedmont, and Cabernet Franc and Merlot, both French varieties but long established in north east Italy. The grapes have very different characteristics but the blend works very well. This Franciacorta has a mid-ruby hue and a complex bouquet of spicy, meaty, grassy and violetty fruit. It has good weight and a mouth-fillingly succulent, fruit-dominated flavour with redcurrants, violets, a touch of cinnamon and a hint of leather on the finish. It makes the lamb taste more tender and more flavoursome while the wine tastes rounder and more fruity.

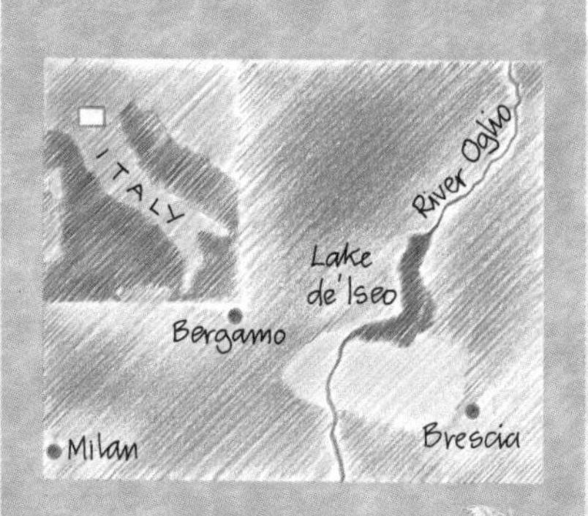

BEST BUYS

Franciacorta Rosso 1987
Contessa Camilla Maggi
Taste guide: medium-bodied

Château Brown 1985
Graves Léognan
Taste guide: medium-bodied

Touraine Cabernet 1987
Domaine des Acacias
Taste guide: medium-bodied

Côtes du Rhône 1987 Les Trois Oratoires
Taste guide: full-bodied

Classic claret

One of the traditional accompaniments to lamb is claret and one which is particularly suitable is **Château Brown 1985 Graves Léognan.** This comes from a district just south of the city of Bordeaux and though better known for its white wines also produces some very distinguished reds. Léognan is a sub-district of Graves and its wines are much admired; 1985 was a notable year in this region. Château Brown is firm, a little lean-tasting but with good fleshy fruit beneath its initial austere impression. Take a mouthful with the lamb and you will find the sweetness of the meat is a perfect foil: a great wine for a perfectly-cooked spring feast.

Free and easy

If you prefer a slightly lighter, zippier wine, try **Touraine Cabernet 1987 Domaine des Acacias.** Touraine is further north than the Bordeaux region and lies along the valley of the river Loire. It is the Cabernet Franc grape which flourishes here and gives a fresh grassiness to the wines. This is noticeable in Domaine des Acacias but it also has good roundness and redcurranty fruit. It makes a great lively partner to lamb, cutting any fattiness the meat may have. It slips down so well that you will probably finish the bottle far sooner than you expect so have one in reserve.

The weather can still be pretty chilly when spring lamb first appears and you may well find yourself turning to warming, comforting wines from further south in France. Some may be too intense and powerful for the comparatively delicate sweetness of lamb, but not all, and certainly not **Côtes du Rhône 1987 Les Trois Oratoires.** It has a most inviting bouquet of damsons, green peppers and black pepper and a soft, rounded but penetrating flavour of rich but spicy sweet-sour fruitiness. It is a great partner for lamb, considerably enhancing its taste, and is excellent value, too.

Any of these wines would be a wonderful accompaniment to lamb cooked in the French way (see below). It is cooked first in a searingly hot oven to brown the meat, then finished at a much lower temperature to tenderise it - rosemary, the classic herb flavouring for lamb, with its delicious, sweet, pungency, simply adds to the flavour combination.

PERFECT PARTNERS

Leg of lamb in the French manner

- ***Preparation:** 10 minutes*
- ***Cooking:** 1½ hours*

1.6-1.8kg/3½-4lb leg of lamb
2 garlic cloves
1tsp rosemary leaves, crushed
juice of ½ lemon
50g/2oz softened butter
salt and freshly ground black pepper

- *Serves 6*

1 Heat the oven to 230C/450F/gas 8. Cut the garlic cloves into slivers. Make small slits all over the lamb with the point of a small, sharp knife, and insert the garlic slivers in the slits.

2 Combine the crushed rosemary leaves with the lemon juice, butter and salt and freshly ground black pepper to taste. Spread the mixture over the lamb and place the meat on a rack in a roasting pan.

3 Place the meat, uncovered, in the oven for 20 minutes. Reduce the heat to 170C/325F/gas 3 and continue to cook for 25-35 minutes per kg/12-15 minutes per lb, or until done.

4 When cooked, transfer the lamb to a heated serving dish and allow to stand for 15 minutes before carving.

NUITS-ST-GEORGES

Let these sensational but reliable reds take the strain out of choosing wines for special occasions

SOMETIMES YOU KNOW that everything, wine as well as food, has to be just right. It may be the first time your in-laws come to lunch, or your work colleagues (or boss!) to dinner, or you are entertaining friends with the reputation of being ace cooks and knowing a thing or two about wine. It's going to be hard enough arranging a well-organized, apparently 'effortless' meal without having to worry about the wine as well. For the centrepiece red, in particular, you need something that will impress and is reliable.

Hang the expense

You may decide that the occasion is important enough to push the boat out – and hang the expense for once. If so, there are few wines more delicious than a really good red Burgundy, and **Nuits-St-Georges 1983 Labouré-Roi** should ensure murmurs of delight at your careful choice. It's not too deep a red – Burgundy rarely is – and has a voluptuous bouquet; almost sweet, it's a little vegetal but far more like raspberries. It's dry, though, and medium-bodied (posh Burgundy is usually elegant rather than 'hearty') with a glorious soft, velvety taste that lingers long after the wine has slipped down. A perfect foil for game, duck or goose, it complements lamb brilliantly too.

If the main course is powerful, or a roast, for example, with strongly flavoured trimmings, a full-bodied wine is called for. **Châteauneuf-du-Pape 1985 Domaine André Brunel** is hard to better. From around Avignon, to the south of the Rhône valley, low-pruned vines on heat-retaining stony soil produce flavour-packed grapes giving powerful, intense wines. Deeply coloured; rich, meaty and earthy on the nose, the taste is weighty, lively and surprisingly fruity, but has a firm, minerally finish that will cut the richness of whatever dish accompanies it. It is excellent, too, for vegetarian dishes where nuts predominate and also if you wanted the same wine with cheese.

Classic claret

If you fear your guests are not that adventurous, it will have to be classic food, with that most classic

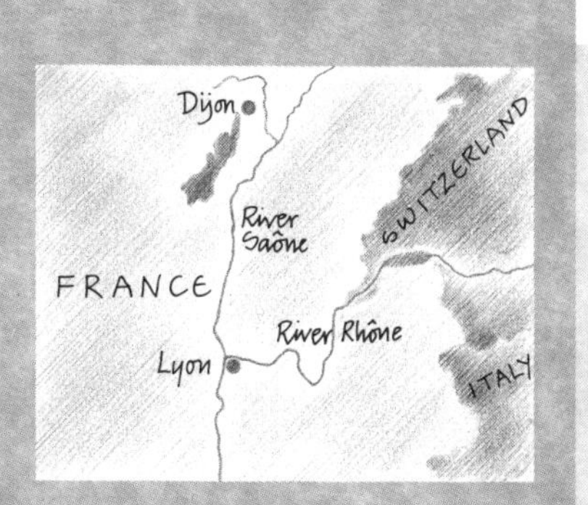

BEST BUYS

Nuits-St-Georges 1983 Labouré-Roi
Taste guide: medium-bodied

Châteauneuf-du-Pape 1985 Domaine André Brunel
Taste guide: full-bodied

Margaux 1986 Vintage Selection
Taste guide: full-bodied

Château La Jaubertie 1986 Bergerac
Taste guide: medium-bodied

Fetzer 1986 California Zinfandel
Taste guide: full-bodied

Chianti Classico 1985 Villa Cafaggio
Taste guide: full-bodied

of smart wines: claret, ie red Bordeaux. For this occasion, though, not just any old claret will do. Go for **Margaux 1986 Vintage Selection**. This doesn't mean it's Château Margaux, one of the top half dozen wines of Bordeaux; instead, it's a wine from the commune of Margaux (from where Château Margaux gets its name). This wine, however, has a pretty smart pedigree itself, being the second wine of one of Bordeaux' *Deuxièmes Crus* – next to the scarce top flight. Deep ruby; with a classy bouquet of cedar and blackcurrants, and an elegant, firm, dry, fruit-and-oak, balanced flavour, no one could but be impressed by your good taste in wines.

Budget buys

If, though, however much you wish to impress, you have to keep to a budget, there are still plenty of wines that will do the trick. Like the superb value **Château La Jaubertie 1986 Bergerac**. Bergerac isn't just the name of a television detective, but a region in France, just next door to Bordeaux and making similar style wines. La Jaubertie has another claim to fame, as the estate is owned, and the wine is made by, Nick Ryman, who sold his countrywide chain of high street stationery stores to retire to a more peaceful existence in the French countryside. His wine gets better every year and the 1986 is a wonderful fruit-filled mouthful with just enough tannin and structure and just the right weight to go well with most traditional British food.

Another way to prove you can find the odd 'snip' and something a little unusual at the same time is to serve **Fetzer 1986 California Zinfandel**. Like most Californian wines there's a full, ripe, softness to the wine that is terrifically appealing. But, unlike many Californian wines, it is not so flavour-packed and concentrated as to outweigh totally any subtlety in your cooking. Zinfandel should get everyone talking too. It's the one grape variety that is claimed as California's own, although it seems to be derived from the little-known and rarely encountered Primitivo from the tip of Italy's heel. Bilberry flavoured, it is far more adaptable to a wide variety of foods than you would think when first tasting it. With casseroles, steaks, poultry, white meat or red; with non-blue cheeses; even with many starters, Fetzer Zinfandel cannot fail.

Classy Chianti

You may feel happier cooking Italian style. An Italian wine would then be the ideal accompaniment. Rather than some exotic name, surprise everyone with Chianti, but a Chianti of such class that will remove forever thoughts of Chianti as a rather simple, possibly even rough wine. The wine is **Chianti Classico 1985 Villa Cafaggio**. This top-quality estate and a superb vintage has turned out a tremendously impressive wine, a choice that will do your cooking proud.

If you are looking to impress your guests at a dinner party, then best end of lamb with cheese and parsley, below, is a great choice. It's quick to prepare (you could get the joints and topping ready the day before, if necessary), and simplicity itself to cook – giving you plenty of time to enjoy both the company of your guests and, of course, the wine!

PERFECT PARTNERS

Best end of lamb with cheese and parsley

- ***Preparation: 15 minutes***
- ***Cooking: 1 hour***

2 best ends of neck of lamb, 6 cutlets each, skinned and chined
salt and pepper
1 tbls mustard powder
3 tbls grated Parmesan cheese
4 tbls fresh white breadcrumbs
3 tbls finely chopped parsley
1 garlic clove, finely chopped
1 tbls red wine
1 tbls olive oil

- ***Serves 4-6*** (!)(££)

1 Heat the oven to 200C/400F/gas 6. Slash the fatty side of the best ends with a sharp knife in a diagonal pattern, season with salt and pepper and put side by side, fatty side upwards, in a roasting tin. Roast for 25 minutes.

2 Place the mustard powder, Parmesan, breadcrumbs, parsley and garlic in a bowl. Add the wine and olive oil and stir.

3 Rub the paste over the fatty sides of the lamb and return to the oven for a further 30-35 minutes, depending on the degree of pinkness of the meat required.

4 Stand the joints upright on a warmed serving dish and serve.

FRAGILE FLAVOURS

BEAUJOLAIS SUPERIEUR 1988

Veal's deliciously subtle flavour can be complemented by a surprisingly versatile range of wines

VEAL IS A favourite meat in a great many countries and should therefore be at home with local wines: having a delicate flavour, it readily takes on the savour of herbs and spices used to accompany it – indeed it can run the risk of being swamped by anything that is too powerful. For this reason, it is wise to avoid any wine that is hefty and full-bodied when you are serving veal. Choose dry white or medium-dry white, rosé, light red or medium-weight red: few foods are as adaptable as this.

Dry white VDQS

If dry wine is your choice, why not go for something good and fruity which is interesting enough in its own right but makes a distinct contrast to the flavour of the meat. One of the nicest is **Côtes de Saint-Mont 1988**. This is classified as a VDQS wine, which means that it is a Vin Délimité de Qualité Supérieure, one step down from the Appellation Contrôlée. Since lots of producers in VDQS areas are trying hard to upgrade their wine to an AC status, many of the VDQS are very good indeed. Côtes de Saint-Mont is situated in south west France in the Gers *département*. It has a fresh fruity aroma with a hint of orange peel and its lively, crisp taste has a depth that makes it linger in the mouth. The flavour comes swiftly through after a mouthful of veal and makes you reach for another taste.

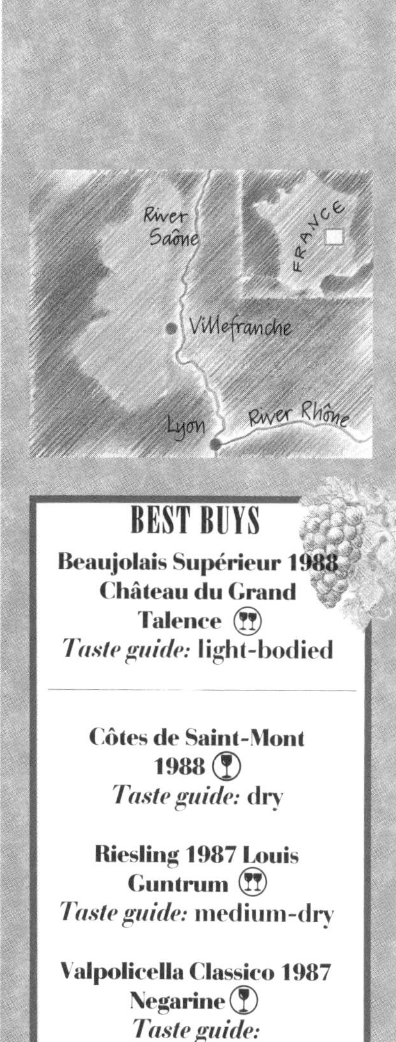

BEST BUYS

Beaujolais Supérieur 1988 Château du Grand Talence
Taste guide: light-bodied

Côtes de Saint-Mont 1988
Taste guide: dry

Riesling 1987 Louis Guntrum
Taste guide: medium-dry

Valpolicella Classico 1987 Negarine
Taste guide: medium-bodied

Light and sweet

One of the frustrating things for the many people who prefer wines that are not too dry is that few of them are at their best with food. Wines that are on the sweet side do not usually enhance a meal – veal, however, contradicts this. A slightly sweet wine has the same effect with veal as has a slightly sweet sauce, based on fruits for example, so that there is an interesting contrast of taste. Be careful not to select a wine that is too sweet, though, since wines taste sweeter with the veal than they do without it. For delicate, not-too-sweet wines, Germany is by far the best place to look. One excellent example is **Riesling 1987 Louis Guntrum:** Riesling is the grape variety and Louis Guntrum a highly regarded producer. The wine is a superb example of the character of the very perfumed, floral Riesling grape – it is medium-sweet but has a good grapey acidity which appears to diminish when drunk with veal.

Everyday Beaujolais

One of the best wines if you want something light and fruity but red is Beaujolais. The popularity of Beaujolais Nouveau may have made people forget that everyday Beaujolais is a delicious wine, readily available at all sorts of prices. It comes from the area north of Lyons and is made from the Gamay grape; it does not need to be especially old to lose the sometimes rasping freshness of the Nouveau variety. Try **Beaujolais Supérieur 1988 Château du Grand Talence:** it tastes just as good with veal as it does on its own with its smoky, sweet-sour raspberry bouquet and its light-weight fruit-dominated taste. It will go well with veal however you cook it and could be the best answer for a dinner party where tastes differ. It satisfies those who feel that meat must have a red wine and it is light enough for those who normally find red wine on the heavy side.

Alla italiana

If you are cooking your veal in an Italian way, and especially if you are cooking it with ham and cheese, you can choose a slightly fuller wine; go for **Valpolicella Classico 1987 Negarine**. The ripe, bitter-cherries bouquet and the rounded cherry flavour of this wonderful wine would be too powerful for plainly cooked veal escalopes, but once the meat is combined with other flavours, as in the rich, and extremely tasty, veal escalopes Parmigiana, below, the wine becomes a great match. The acidity of this Italian wine is a perfect foil for cheese - and there are two quite different cheeses in this recipe, one, mild and creamy Mozzarella, the other, strong and pungent Parmesan - making this wine the perfect choice.

PERFECT PARTNERS

Veal escalopes Parmigiana

- ***Preparation: 30 minutes***
- ***Cooking: 14 minutes***

4 – 100-150g/4-5oz veal escalopes, at room temperature
oil for deep frying
2tbls flour
salt and freshly ground black pepper
1 egg, lightly beaten
4tbls thin cream
25g/1oz fresh white breadcrumbs
8tbls freshly grated Parmesan cheese
4 skinned tomatoes
100g/4oz Mozzarella cheese, thinly sliced
2tbls finely chopped parsley
watercress, to garnish

- ***Serves 4***

1 Heat the oven to 230C/450F/gas 8 and heat the oil in a deep-fat frier to 190C/375F. Place the veal escalopes between sheets of stretch-wrap and flatten until thin with a meat bat or rolling pin.

2 Sprinkle the flour onto a plate and season to taste with salt and pepper. In a shallow bowl, combine the egg and cream and mix to blend with a fork. Sprinkle the breadcrumbs into a separate shallow dish and mix with half the grated Parmesan cheese.

3 Toss each escalope in seasoned flour to coat, then in the egg and cream mixture, draining off any excess. Coat in the breadcrumbs and cheese mixture, patting firmly with your hands. Lay the prepared escalopes flat on a tray.

4 Deep fry 2 escalopes at a time, for 30 seconds on each side. Drain well on absorbent paper and place in a large ovenproof dish. Spread the sliced tomatoes over each one and sprinkle with the remaining grated Parmesan cheese. Top with thin slices of Mozzarella cheese and sprinkle with the parsley.

5 Bake in the oven for 10 minutes or until the cheese is melted and golden brown. Decorate with watercress.

FORCEFUL FLAVOURS

COTES DE BEAUNE

Inexpensive, tender but strongly flavoured, liver requires specific types of wine to match

BEST BUYS

Côtes de Beaune
Taste guide: medium-bodied

Tres Torres Sangredetoro 1987
Taste guide: full-bodied

Syrah: Vin de Pays des Coteaux de l'Ardèche
Taste guide: medium-bodied

California Zinfandel Monterey County, Taylor California Cellars
Taste guide: medium-bodied

Cahors 1985 Domaine de Lerat-Montpezat
Taste guide: full-bodied

FINELY CUT SLICES of liver, lightly cooked so they are still just about pink in the middle, tender as butter and gently, but strongly flavoured, is one of the most underrated dishes. Liver is still comparatively inexpensive too, so, even with a decent bottle of wine, the meal shouldn't break the bank.

Beautiful Burgundy

The taste of liver is quite gamey and so wines that would go well with strong game are a good bet with liver too. One of the most exciting combinations with game is red Burgundy, from the Pinot Noir grape variety and, indeed, it makes just as stimulating a match with liver. For a great match that isn't too pricy, it would be hard to find better than **Bourgogne Hautes Côtes de Beaune 1986.** Wine from the Hautes Côtes of Beaune is wine from the higher slopes, further back from the Saône river. Many of the most sought-after wines come from the lower slopes, leaving the Hautes Côtes free from the excessive demand that forces up prices. Yet, as this wine shows, with its clear, bright ruby colour, its alluring ripe, gamey, raspberry taste, the Hautes Côtes can produce some delicious wines too. The gaminess on the nose complements the gamey aromas of the liver perfectly.

Spanish support

Liver can be well matched by a wine with plenty of rich fruit such as **Tres Torres Sangredetoro 1987**. Its origin is Spain, from the north-east near Barcelona: the region is called Penedes. It comes from the remarkable estate of Torres, which has done more than any other to make fine quality Spanish wines and put them on the map. This wine is made solely from the traditional Spanish grape varieties Garnacha and Cariñena.

Peppery flavours

Another category of wines that has an affinity with liver is the type with a rather peppery bouquet and taste. These sorts of wines usually come from around the southern part of the Rhône valley and from grape varieties such as Syrah, Grenache and Mourvèdre. Syrah is the grape that produces the classiest wines and is being used more and more, even for the uncomplicated Vins de Pays. **Syrah, Vin de Pays des Coteaux**

de l'Ardèche is an excellent example. It has an enticingly rounded bouquet with aromas of plums, blackberries and cherries as well as the nose-tickler of black pepper. These fruits come through on the taste too, with an even stronger pepperiness – that is, until the wine is drunk with liver. The liver removes the peppery sensation and 'sets the fruit free' so that the wine tastes considerably richer, fruitier – and better.

Zinfandel is California's only native grape variety, all the others having been brought as cuttings from France. It seems to be the same variety, however, as the Primitivo – which comes from the heel of Italy, although no-one is sure how it found its way to California. It is surprising for the number of styles of wine it can produce. It can make the lightest of barely-pink rosés, or blush wines as they are often called, or wines so big, full, dark and rich that they are almost like Port. **California Zinfandel Monterey County** bottled by Taylor California Cellars is just the right level for liver. It is mid-weight with lively raw fruits – loganberries and not quite ripe blackcurrants, and a liquorice tang. With liver, though, the tanginess disappears, the wine fleshes out and the crisp fruitness is a good complement to the liver.

Another good choice which has all the liver-matching characteristics - ripe but raw fruits, a peppery, liquorice tang, and an affinity with game - is **Cahors 1985 Domaine de Lerat-Monpezat**. Cahors lies in south west France, around the valley of the Lot river, and the wine is made from the Malbec grape variety together with some rounded Merlot and tannic Tannat. It is quite moreish with a number of dishes, but with liver it really is in its element. The tannin softens right down, the wine appears rounder and the fruits softer. It might be a touch too full-bodied if you are cooking calves' liver in preference to the slightly stronger lamb's liver. In this case go for the lighter, but extremely fruity Zinfandel, which with its lively blackcurrant flavour would be a delicious match, or try Côtes de Beaune with its lovely raspberry taste.

PERFECT PARTNERS

Calves' liver with mushrooms

- ***Preparation: 15 minutes***
- ***Cooking: 15 minutes***

16 – 3mm/⅛ in thick small slices calves' liver, about 450g/1lb weight
3tbls flour
salt and freshly ground black pepper
225g/8oz button mushrooms
25g/1oz butter, plus extra, if necessary
2tbls olive oil, plus extra, if necessary
lemon juice
For the garnish:
5 thin twists lemon
5 small sprigs of watercress

- *Serves 4*

1 Season the flour with ½tsp salt and a good sprinkling of freshly ground black pepper, mixing them well together on a plate before cooking the liver.

2 Wash or wipe the button mushrooms clean. Trin the stalks and cut the mushrooms into thin slices.

3 In a heavy frying pan, heat half the butter and oil with 1tbls lemon juice. Add the sliced mushrooms. Toss until well coated in the lemony fat and sauté over a moderate heat for 5 minutes, or until soft, stirring occasionally with a wooden spoon. Season to taste with salt and freshly ground black pepper. Keep hot.

4 Dry the liver on absorbent paper. Coat each slice with seasoned flour, shaking off any excess. In a large frying pan, heat the remaining butter and olive oil. Add half the liver slices to the sizzling fat in a single layer. Sauté on one side for 1 minute, or until droplets of blood appear on the uncooked surface.

5 With a spatula, turn the slices over and sauté for 1 minute on the other side. The liver should be nicely browned on the surface but still faintly pink inside. Transfer to the heated serving dish with the mushrooms and keep warm while sautéeing the remaining liver using more butter and oil if necessary. Garnish with lemon twists and watercress.

LA TUQUE BEL-AIR

Red meat with red wine is the general rule, but some are better than others. Choose a wine with lots of tannin and you won't go wrong

WHEN THE DIGESTIVE juices start working overtime at the prospect of good, succulent red meat, be it as a steak, a roast or in a casserole, the wine to go with it needs to be pretty robust, too. Rich, ripe, fruity reds which may seem really impressive at first taste can often fade into insignificance when partnering big and flavoursome carnivorous dishes.

A taste for tannin

Often the best red wines for red meats are those which seem a bit harsh on their own, those with noticeable tannin. Tannin is the substance in wine, usually only found in reds, that makes the inside of your mouth feel dry – like drinking tea without milk. It is found in grape skins (try chewing a grape skin and you will feel the tannin on your teeth) and it is extracted from them along with the colour in red wine making.

Though tannin makes the wines a bit off-putting when drunk on their own, it is an important ingredient of wines that will age. When wines with tannin meet red meat protein, though, they really come into their own. The tannin (and the wine's acidity) help tenderize the meat in the mouth, so it tastes even better than ever, while the meat protein removes the tannin so the wine tastes soft, silky smooth and utterly delicious.

BEST BUYS

Domaine la Tuque Bel-Air 1985
***Taste guide:* full bodied**

Domaine de Diusse 1985 Madiran
***Taste guide:* full bodied**

Tinto da Anfora 1983
***Taste guide:* medium-bodied**

Bourgogne Pinot Noir 1985
***Taste guide:* medium-bodied**

Starry eyed

The classic example of a starred partnership between red wine and red meat is steak with claret, the red wine of Bordeaux. Claret can be horrendously expensive, with names like Lafite, Latour and Pétrus fetching unbelievably high figures. It can also be very reasonable, often sold under a merchant's own-label just as claret or Bordeaux. In between there are all levels of quality and price. The simplest wines are lighter and less concentrated than the rest and may be just OK, no more, with that steak.

Heat and dust

Bordeaux is often reasonable value, but, surprisingly, good value is rare. It is found in **Domaine la Tuque Bel-Air 1985 Bordeaux Superieur-Côtes de Castillon.** Bordeaux is a large wine region, so is split into smaller districts, each with its own character. Côtes de Castillon is one of the more outlying districts, towards the east of the region, which rarely produces the heftiest wines. But 1985 was a year of great heat and dryness, giving small berries, and therefore a greater proportion of skins to juices.

A deep ruby red colour, this wine has a rich fruit cake and spice bouquet which is typical of the Merlot grape from which it is made (together with Cabernet Sauvignon and Cabernet Franc). It's firmly flavoured, with noticeable tannin and good acidity, but not lacking in strong, fleshy, classy fruit that will shine through when drunk at lunch or dinner.

Big with beef

Bigger, chunkier and more tannic is **Domaine de Diusse 1985 Madiran.** From the same year, 1985, it also comes from further south west in France, the zone of Madiran, towards the Pyrenees. The main reason for the wine's uncompromising style, though, is its predominant grape variety, Tannat, which gives depth – in colour, smell and flavour.

A big brambly bouquet leads to masses of taste; plenty of fruit but almost aggressive structure. Certainly not a wine for attempting on its own, but just try it with a rich, beefy casserole and taste how both are immediately enlivened.

Rich with roasts

Not often do wines for meat dishes actually smell meaty. **Tinto da Anfora 1983** does. It is one of the best examples of the skill that has been used in Portugal to make excellent red wines. After being vinified the wine was left until the spring of the following year when it was put into chestnut casks where it matured for a year before bottling. A bright, mid-depth red, it smells plummy as well as meaty.

Its taste is surprisingly original. The predominant fruit seems almost sweet, and is softened and rounded by the gentle wood flavours. The tannin isn't noticeable at first, but appears later and the wine isn't too full bodied. It comes into its own with simply roasted joints of meat.

Game, set and match

Burgundy is the exception that proves the rule. It isn't necessarily very tannic and it is lighter bodied than many reds, yet it has a wonderful affinity for game, the most powerfully flavoured fare. Burgundy is often very expensive. It is also not always easy to gauge the quality of the wine from its price.

Sometimes the simplest wines give the most pleasure, as is the case with **Bourgogne Pinot Noir 1985 Groupement de Producteurs Buxy.** Pinot Noir is the sole grape variety for nearly all red Burgundy and it gives a glorious raspberry-sweet, sometimes almost vegetably, aroma to the wine. The co-operative of producers at Buxy has encapsulated the style of this grape, and, by ageing the wine in oak, has produced a deliciously elegant, rounded, fruity Burgundy that is packed full of flavour.

PERFECT PARTNERS

Roast wing of beef

- ***Preparation: 10 minutes, plus standing***
- ***Cooking: 1½-2½ hours, plus standing***

2.3kg/5lb wing rib of beef, on the bone
salt and pepper
3tbls Dijon mustard
1tbls dried oregano
2tbls flour
300ml/½pt beef or other stock

- ***Serves 6-8***

1 Remove the beef from the refrigerator at least 2 hours before cooking and leave to come to room temperature.

2 Heat the oven to 200C/400F/gas 6. Sprinkle the beef with pepper. Mix together the mustard and oregano.

3 Place the joint in the oven and cook, basting and turning as necessary. Allow 33 minutes per 1kg/15 minutes per lb, plus an extra 15 minutes, if you like your beef rare. If you prefer your meat well done, allow up to 55 minutes per kg/25 minutes per lb, plus 20 minutes extra Thirty minutes before the meat is done, spread the mustard mixture over the fat.

4 Remove the beef from the oven, transfer to a carving board and stand in a warm place for 15-20 minutes.

5 Meanwhile, strain any excess fat from the roasting tin, leaving about 2tbls of the juices. Add the flour and stir over low heat for 3-4 minutes. Gradually add the stock, stirring, and bring to the boil. Season and simmer for 5 minutes.

6 To carve the beef, rest the joint on its rib bones. Run a sharp knife down the backbone and then along the ribs to free the meat from the bones. Remove the meat from the bones and place it, fat side up, on the carving board. Carve downwards into thin slices and arrange the slices on a heated serving dish. Pour the gravy into a heated sauceboat and hand it round separately.

CROZES HERMITAGE

If you're undecided about a red wine, then buy a half bottle. That way you won't make an expensive mistake and you can afford to be more adventurous in your choice

A HALF BOTTLE IS a very convenient size for certain occasions. It is ideal if there are just two of you and you don't want to drink too much. It is perfect if there are up to four of you as it gives you the chance to have a different wine with each course, or try two wines with the same dish. It also means you can drink more expensive wines than you would otherwise be able (or willing) to afford. It is even helpful as a trial purchase of a wine you are not sure whether you are going to like or not before you buy a larger quantity.

Going halves

It is surprisingly difficult to find half bottles on the shelves. There are various possible explanations for this. Halves are not popular with many importers who believe, rightly or wrongly, that producers always put the dregs of a bottling run into halves, or at least not their best wine. They are not popular with producers as they are more fiddly to bottle and label and the wine matures faster in a smaller container. Neither are they particularly cost-effective as they still require a bottle (albeit smaller), a label, a cork and a capsule, and they are proportionately heavier to ship. Nevertheless, they are so useful you would have thought that these minor disadvantages would have been overcome by most merchants. Unfortunately, this is not so and the search for half bottles is often a major hunt, though well worth while.

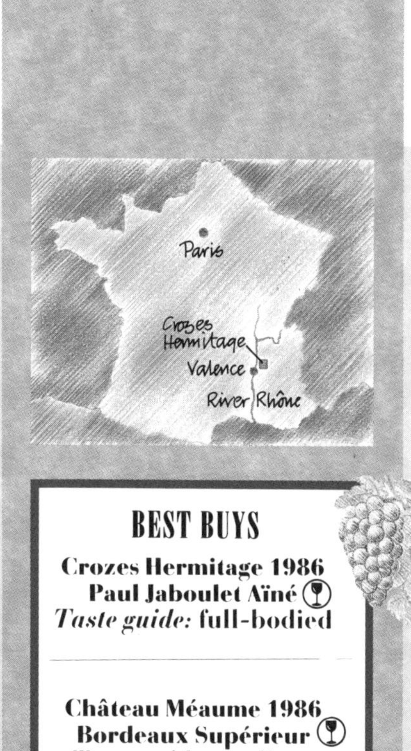

BEST BUYS

Crozes Hermitage 1986 Paul Jaboulet Aîné
Taste guide: **full-bodied**

Château Méaume 1986 Bordeaux Supérieur
Taste guide: **medium-bodied**

St Estèphe 1985
Taste guide: **medium-bodied**

Capitel Monte Fontana 1984
Taste guide: **full-bodied**

Brilliant Bordeaux

For the red halves the most likely source of supply is the renowned French region of Bordeaux. A number of the châteaux there have traditionally always bottled at least some of their production in halves and the custom seems to have lingered. As an example, there's **Château Méaume 1986 Bordeaux Supérieur.** The complexity of flavour found in most Bordeaux reds comes from the blend of three grape varieties used. This is admirably demonstrated in Château Méaume with the blackcurranty character of Cabernet Sauvignon, the grassiness of Cabernet Franc and the soft fleshiness of Merlot all apparent. Still quite firm, with plenty of tannin, it needs good roast meat or poultry, a hearty casserole, steak and kidney or just offal to show it at its best. It will keep for several years if you don't feel like drinking it straight away.

Also from Bordeaux is **St Estèphe 1985** from Caves et Entrepots de Malecot. St Estèphe is a small area in the part of Bordeaux known as Médoc, a little north of the town itself. It is one of the six or seven communes where most of the best and most expensive wines, the Crus Classés, come from. Indeed, there is hardly any wine from St Estèphe that isn't of Cru Classé status and it is rare to see a label simply declaring 'St Estèphe' without some exalted château name too. This wine, however, does indeed come from one of these posh châteaux and, though good, is just considered not quite good enough to be sold under the château label. With a preponderance of Cabernet Sauvignon, it has an inviting cedary character from the oak barrels in which it has been matured. A good attack of clean and complex flavour, with fruit and tannin well balanced, it would enhance game, steaks or any grilled food.

Red from the Rhône

Crozes Hermitage 1986 Paul Jaboulet Aïné comes from the northern Rhône valley. Part of this valley is overshadowed by the impressive, steep Hermitage hill, completely vine covered. The back slope of the hill is more gentle and is where Crozes Hermitage is produced. From Paul Jaboulet Aïné, one of the Rhône's foremost producers, this is a terrific, full-bodied wine, redolent of blackcurrant fruit gums, tar and spice, which are typical of the powerfully flavoured Syrah grape variety, beautifully balanced and with a long-lasting taste. Enjoy it with any red meat or non-blue cheeses.

Sweet and strong

It's unusual to find a sweet red wine but Recioto della Valpolicella is very special. It is made from bright and healthy grapes for Valpolicella, which are dried (and thereby concentrated) for three months before pressing and slowly fermenting. When the yeasts stop working there's still a little grape sugar left, so the wine is somewhat sweet. Tedeschi are masters at the art and their **Capitel Monte Fontana 1984** is a prime example of Recioto. Deep pinky purple, it smells rich and porty, yet still reminiscent of Valpolicella. Big, rich, sweet and strong, without any heaviness, it is a delight at the end of a meal with cheese, nuts, raisins, cakes, or on its own. You only need a small glass, so a half bottle goes a long way.

Of course, when you are dining *à deux*, then a half bottle makes a lot of sense. And what better to eat with your choice of red than steak with red wine sauce (see below). Despite the name, the sauce uses only 50ml/2fl oz of wine, so there will still be plenty left over from cooking to give you two decent-sized glassfuls to drink!

PERFECT PARTNERS

Steak with red wine sauce

- ***Preparation: 15 minutes***
- ***Cooking: 20 minutes***

2 × 175g/6oz entrecôte steaks
For the sauce:
1½tbls unsalted butter
2 shallots, finely chopped
50ml/2fl oz red wine
125ml/4fl oz beef stock
½tsp finely chopped fresh thyme or ¼tsp dried thyme
1 small bay leaf
25g/1oz beef marrow, sliced
½tsp lemon juice, to taste
salt and pepper

- ***Serves 2***

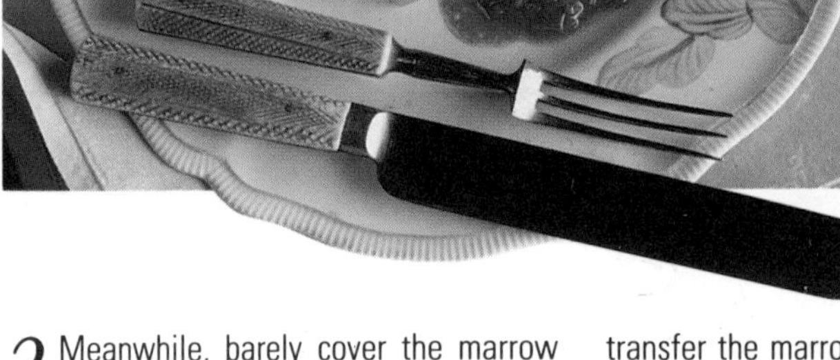

1 Cook the sauce while you grill the steaks. Melt half the butter in a small saucepan over medium heat and cook the shallots, stirring constantly, for 2 minutes. Add the wine to the pan and boil for about 2-3 minutes or until the liquid is reduced by half. Add the beef stock, thyme and bay leaf and simmer the sauce for 5 minutes.

2 Meanwhile, barely cover the marrow with water and poach gently for 2 minutes.

3 Remove the sauce from the heat and remove the bay leaf. Stir in the remaining butter. With a slotted spoon, transfer the marrow to the sauce and add the lemon juice and salt and pepper.

4 Pour into a warmed sauceboat. Transfer the grilled steaks to a warmed serving dish. Spoon a little of the sauce over the steaks and serve.

CABERNET SAUVIGNON VIN DE PAYS D'OC

Nothing enhances a good steak more than a bottle of excellent red wine

ALTHOUGH OFTEN OVERLOOKED in the search for novelty in meals, there are few things better than a good steak. It is also one of the best foods with which to drink red wine. For the acids in wine help to tenderise it, making it softer and more digestible. For its part, the steak removes the tannin in wine, so making it taste much rounder and more enjoyable.

Add to the natural benefits of wine and steak together the benefits of a wine with a taste that harmonises well with steak, and you have a wonderful partnership indeed. And there are plenty of such good matches, at all price levels.

Southern superiority

For really good value and a brilliant partnership with steak, there are few better wines than **Cabernet Sauvignon Vin de Pays d'Oc**. Cabernet Sauvignon is one of the world's most famous grape varieties: its style changes with the climate and soil of wherever it is being grown. Cabernet Sauvignon in Bordeaux, its major home, can be quite austere, but further south in the Languedoc-Roussillon region of south west France, known as the Oc, it is softer and rounder. It has gloriously rich, blackcurranty fruit, with just a little tannin. With steak, that tannin is no longer noticeable, but the wine seems more incisive and classier, while the steak gains succulence and flavour.

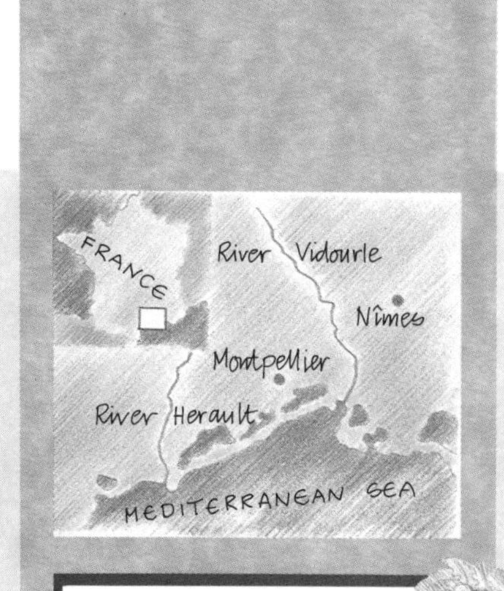

BEST BUYS

Cabernet Sauvignon Vin de Pays d'Oc
Taste guide: **medium-bodied**

Saint Joseph 1986 Cuvée du Bicentenaire – Les Producteurs de Saint-Désirat
Taste guide: **full-bodied**

Moulin-à-Vent des Hospices 1985
Taste guide: **medium-bodied**

Orlando South East Australian 1986 RF Cabernet Sauvignon
Taste guide: **full-bodied**

Château des Gondats 1987 Bordeaux Supérieur
Taste guide: **full-bodied**

Support from Syrah

Another wine that turns a steak meal into something really exciting is **Saint Joseph 1986 Cuvée du Bicentenaire** from Les Producteurs de Saint-Desirat. Saint Joseph comes from France's Rhône valley. Most of the wines from along the Rhône, such as Côtes du Rhône and Chateauneuf de Pape, come from the southern part of the valley, around Avignon, and are made from a blend of any number of grape varieties. Saint Joseph lies a little further north, where the river sweeps down through a narrow gorge and the mistral wind often batters the hillsides. In this part only one variety is grown, staked hard to the ground to secure it against the wind: the Syrah. Syrah has a remarkable bouquet. Intense and powerful, it is reminiscent of blackcurrant fruit gums with spice. On some wines it can be almost overwhelming, but on this Saint Joseph the balance is just right. There's plenty of body and a strong, peppery and liquorice flavour. This is just the right flavouring to act as a counterpoise to the meat.

Windmill magic

Beaujolais is usually light, fresh and cheeringly fruity – a great wine for many occasions but not one of the best matches for steak. There are, however, a number of villages within the Beaujolais area which produce bigger, weightier wines. The village said to be the best of all is Moulin-à-Vent. It is also practically the only village whose Beaujolais will age well, becoming more mellow and occasionally getting rather like the red Burgundy which comes from further north. Aged Moulin-à-Vent, then, is a good contender for drinking with steak. Unfortunately, it is not often seen and is rather dear. It is fortunate that **Moulin-à-Vent des Hospices 1985** is available. With its attractive smoky raspberries bouquet and its lightish, but velvety and harmonious taste, enlivened by good balancing acidity, it is a good wine for steak.

Another example of the Cabernet Sauvignon grape is **Orlando South East Australian 1986 RF Cabernet Sauvignon**. It has the warmth and ripeness of the south east corner of Australia where it was grown. A deep pinky-purple, you can almost smell the ripeness bursting out of it, with its rich, brambly bouquet. Luscious, full and velvety, it packs in a lot of comfortingly rounded, seemingly slightly sweet flavour. With the steak, however, it tastes firmer, dryer and more serious, bringing out the steak's flavour and making it taste more succulent.

One of the classic accompaniments to steak is red Bordeaux. **Château des Gondats 1987 Bordeaux Supérieur** shows why. Although its official category is Bordeaux Supérieur, only one notch up from straightforward Bordeaux, the wine has a much better pedigree than this would imply. In fact, it is the 'second wine' of a very grand château, that of Marquis de Terme. Like most of the top Bordeaux Châteaux, Marquis de Terme makes a very strict selection of the wine it bottles under that name. Anything almost but not quite top notch is bottled as Château des Gondats – for a much lower price. So it is a real bargain. It has a delicious, classy, blackcurrant and cedary bouquet and a refined, elegant, but quite tannic taste. This wine is much better when it has mixed with the air, so open it well in advance, or better still, decant it.

Grilled steaks with Roquefort butter (see below) are a delicious partner to any of these red wines. If you find Roquefort too strong, then make the butter in another flavour, such as watercress (omit the cheese and substitute 3tbls finely chopped watercress for the parsley), or parsley (omit the cheese and add 1tbsp lemon juice and ¼tsp paprika).

PERFECT PARTNERS

Grilled steaks with Roquefort butter

• ***Preparation: 10 minutes***

• ***Cooking: 8 minutes***

4 rump steaks, weighing about 175g/6oz each, 20mm/¾in thick
salt and black pepper
olive oil
fresh salad vegetables to garnish
For the Roquefort butter:
50g/2oz butter
1tsp flour
25g/1oz Roquefort cheese, crumbled
2tbls finely chopped parsley
lemon juice

• ***Serves 4***

1 Take the steaks out of the refrigerator to bring to room temperature and wipe with a clean damp cloth. Trim off excess fat, then nick the remaining fat around the edge (Masterclass, page 203). Beat the steaks on both sides with a wooden rolling pin, or meat bat, to tenderize the meat and season with pepper.

2 To make the Roquefort butter, mash the butter and flour in a small bowl with a wooden spoon. When smooth, add the crumbled Roquefort cheese and the finely chopped parsley and mash again. Add lemon juice to taste and place the mixture on a piece of greaseproof paper. Pat the mixture into an oblong, wrap with the paper and chill in the refrigerator.

3 Heat the grill to high, oil the grid of the grill pan. Season the steaks again with salt and pepper. Place the steaks on the grid and grill the meat 7.5cm/3in from the heat for 2 minutes on each side for rare, 3 minutes for medium rare and 4 minutes each side for well done.

4 Slice the Roquefort butter into neat pats. (For other flavoured butters, see above). Transfer the steaks to a heated serving platter and serve immediately acompanied by a mixture of seasonal salad vegetables.

MONTEPULCIANO

There's something about a glass of smooth red wine that warms and cheers, and helps soothe away niggling daily worries

THE PROSPECT OF a glass of any wine is cheering in itself but when you are feeling particularly tired, overwrought or just fed up there are some wines that are just so much better at helping the misery disperse than others. You don't want anything too light or too heavy. A serious wine for 'appreciation' is out as is anything which is too frivolous or celebratory.

Reds are in the forefront of the cheer-up brigade. The right ones for the job have a particular type of seemingly ebullient fruitiness that almost bounces out of the glass at you and jauntily fills the mouth. You don't have to worry about these sorts of wines; you know the last mouthful will taste as good as the first. You know that they will perk you up on their own but are also

River Tiber
River Pescara
ABRUZZO
Rome
ITALY
Naples

BEST BUYS

Montepulciano d'Abruzzo
Taste guide: **medium-bodied**
Serve with: **chicken, duck, cold meats, salami**

Quinta d'Abrigada 1984
Taste guide: **medium-bodied**
Serve with: **rice dishes, food in batter or breadcrumbs**

Crozes-Hermitage 1986
Taste guide: **medium-bodied**
Serve with: **steak, lamb chops, casseroled beef**

Le Carla 1986 Fitou
Taste guide: **medium-bodied**
Serve with: **quiches, white meat in light sauces**

DECANTING

You shouldn't need to decant wine that often. The only time it is essential to decant is if a wine has a sediment – and that only happens with fine red wines that have been aged for a number of years. If you do have such a bottle make sure it is stored lying down to keep the cork moist and avoid air getting in; slowly turn it upright a day before you want to drink it.

Remove the capsule and cork without jolting the bottle too much, get a torch, then carefully lift the bottle over the decanter. Any jug will do if you haven't the real thing. Pour the wine as gently as you can into the decanter: your aim is to disturb the deposit as little as possible. Keep the torch underneath the bottle near its neck so you can see through the wine. After pouring most of the contents you will see the first bits of sediment begin to float towards the neck. As soon as the sediment is about to flow out, stop pouring. The dregs can be poured into a spare glass, left to settle and used for cooking.

resilient enough to be genial partners to a wide range of possible meals. In short they are just what the doctor ordered.

The right stuff

An example is **Quinta d'Abrigada 1984** from Portugal. Portugal offers many up-front, enlivening, fruit-filled red wines. Its local grape varieties and its climate, hot in summer and cold in winter, are ideal for producing punchy, cheering styles. Quinta d'Abrigada is simply a mouth-filling, medium-bodied, plummy, warming, slightly spicy, rounded wine. What makes it so deliciously moreish, though, is the succulent quality of its fruit and its balanced structure: neither hard nor jammy.

Italy also has the climate for hearty, cheering reds. Often, though, Italian wines are far too firm, too concentrated, or too light for the rôle, or just best suited to more serious meals or occasions. This is because many Italian grape varieties naturally produce such wines and because too few Italians drink wine on its own to appreciate our needs. Not all Italian reds, though, are outside the category, thank goodness. Montepulciano is a grape that can brighten the most dismal day and reaches its peak in the central southern Adriatic region of Abruzzo. **Montepulciano d'Abruzzo** produced by AVUR is crisp, like English blackberries, meaty, and with an exuberance of fruit flavour that wakens, enlivens and truly cheers.

Another grape that can fit the bill is Syrah, even though it sometimes produces big, chunky, concentrated wines: good in themselves but not perky enough for cheering up. Not **Crozes-Hermitage 1986** from Délas, though; it is as lively as one could wish. The steep hill of Hermitage, traditionally reputed for impressive, long-lived wines, overlooks France's Rhône valley and is covered with Syrah vines. Crozes-Hermitage, on the other hand, comes from the gentler slope of the hill, on the side away from the river, where the Syrah grape still displays its blackcurrant-fruit-gums character but in a less obvious way.

The gentle touch

A gentler wine, which will coaxingly rather than punchily brighten your day, is from Fitou in the extreme south west of France, near Narbonne. **Le Carla 1986 Fitou** from the Vignerons de Resplandy is named after a place of pilgrimage in the mountains there. It is a mixture of Grenache and Carignan grapes which together have made a lightish-weight, herby, plummy, vinous, warm wine with a soft smoky, tarry bouquet, which is very easy to drink, with no rough edges at all. It can certainly comfort and will almost definitely soothe away the problems and the harassments of even the most tense and fraught day.

PERFECT PARTNERS

Minute steaks with mushrooms

- ***Preparation: 5 minutes***
- ***Cooking: 10 minutes***

4 minute steaks, each weighing 200g/7oz
1 large garlic clove, halved
salt and pepper
1tbls oil
50g/2oz butter
225g/8oz button mushrooms, thinly sliced
juice of ½ lemon
1tbls chopped parsley or snipped chives, to garnish

- ***Serves 4***

1 Rub the steaks with the cut side of the garlic clove and season with pepper.

2 Heat the oil and half the butter in a heavy frying pan. When the foaming subsides, increase the heat slightly and fry the steaks for 2-3 minutes on each side, for medium-done steaks.

3 Transfer to a warmed serving dish, season with salt and keep hot.

4 Melt the remaining butter in the pan and fry the mushrooms over low heat for 3 minutes.

5 Increase the heat and sprinkle the mushrooms with the lemon juice. Cook for 3 minutes or until most of the moisture has evaporated, stirring constantly.

6 Spoon the mushrooms over the steaks and serve immediately, garnished with either chopped parsley or snipped chives.

CHATEAU LEON 1985

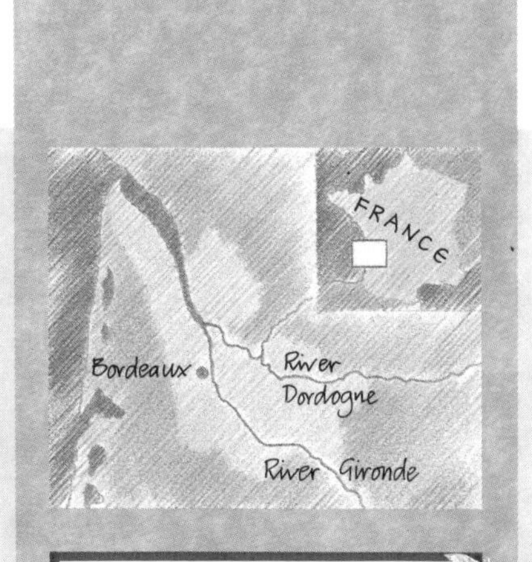

BEST BUYS

Château Léon 1985 Premières Côtes de Bordeaux
***Taste guide:* medium-bodied**

El Coto 1985 Reserva Rioja
***Taste guide:* medium-bodied**

Orlando 1986 RF Cabernet Sauvignon
***Taste guide:* full-bodied**

Côtes du Rhône 1987 Château du Bois de la Garde
***Taste guide:* medium-bodied**

Côtes de Nuits-Villages 1983 Labouré Roi
***Taste guide:* medium-bodied**

Bourgogne Pinot Noir 1986 Groupement de Producteurs Buxy
***Taste guide:* medium-bodied**

Easy choices for dinner party red wines

IT IS NORMALLY a little easier taking a red wine to a dinner party than a white. You don't have to worry too much about its temperature as you know there will be time for it to warm up before you start to drink it. On the other hand, you don't know whether it will be drunk at all. Your hosts – or other guests – may already have provided enough red. So don't take along a particularly prized bottle that you are keen to try, for you may find it goes straight into your hosts' 'cellar' and you never see it again.

If you are lucky enough to have a particularly fine or old bottle, don't take that either as it may well have a sediment which will get churned up on the journey. Save it for when folk come round to you. Anyway, the essence of a dinner party wine is one that can be enjoyed by everyone who might be there, regardless of their taste, and one that goes with a wide variety of dishes.

Bordeaux

It is hard to go wrong with reds from Bordeaux. The British have been drinking and enjoying them for over eight hundred years. Not only is the taste both refined and appealing, but the wines go perfectly with our diet. Red Bordeaux, often called Claret, is made from a mixture of Cabernet Sauvignon, Cabernet Franc and Merlot grapes. Some wines have more of one grape type, some more of another. This means there is scope for endless variety on the theme and there is seemingly no limit to the number of really great wines to be tried. There is no need to spend a fortune to enjoy one – although some bottles are incredibly expensive. Just buy **Château Léon 1985 Premières Côtes de Bordeaux.** A deep, bright ruby colour, with full, firm, flavours. Try it with practically any meat-based main course such as casseroles, meat pies or roasts, as well as pâtés, terrines and gentle cheeses.

Rioja

One wine that always goes down well is Rioja. Even some people who rarely drink red wine enjoy it. So it is a good bet to take to a dinner party when you don't know who your fellow guests are. Rioja comes from north east Spain, from the Ebro valley. Its style can vary depending on which part of the area the major part of its grapes come from, and on how long it is matured before bottling. For a good oaky taste, which is Rioja's hallmark, you need a bottle which has had at least one year in oak barrels. This is compulsory for wines called Reserva, as is three years' ageing in total to ensure they are mellow. **El Coto 1985 Reserva Rioja** is a good example with its bright ruby colour. Drink with roast and grilled red meats and medium-weight casseroles.

Down Under

Australia is another area that has won over most wine drinkers. For a dinner party Australians go for the Cabernet Sauvignon grape which is the most versatile with food. You will find it hard to beat **Orlando 1986 RF Cabernet Sauvignon** from south eastern Australia. Deep, big, rich, ripe and herbily blackcurranty-fruity, it has most cheering of taste.

Côtes du Rhône is made from so many different grape varieties and in so many different ways, it is hard to know what any bottle will taste like. Sometimes it is almost as light and fruity as Beaujolais, sometimes big, powerfully full-bodied and tannic. Neither of these styles is useful for a dinner party as the food may overpower the wine or be overpowered. One of the nicest and most acceptable wines to take is **Côtes du Rhône 1987 Château du Bois de la Garde.**

Burgundy

If you want something that will impress, take a bottle of Burgundy. Find a **Côte de Nuits-Villages 1983 Labouré Roi,** which is a terrific buy as it is well matured and tastes delicious – but is a good price.

If, though, you want Burgundy-on-a-budget, then there is none better than **Bourgonne Pinot Noir 1986 Groupement de Producteurs Buxy.** It is also perfect for those not-so-grand evenings when you are cooking at home for an informal supper - when the mood is relaxed and the food is easy-going, hearty fare such as casserole of beef (see below). What's perfect about this dish, apart from the superb flavour, is that it can be left to cook slowly in the oven, so you have time to spend with your guests. Lamb, cooked slowly in the same way, would be equally good.

PERFECT PARTNERS

Casserole of beef

- ***Preparation: 30 minutes***
- ***Cooking: 2½ hours***

2 Spanish onions, finely chopped
4 garlic cloves, finely chopped
3 shallots, finely chopped
275g/10oz piece of green bacon or ham, cut into strips
25g/1oz butter or 2tbls chicken fat
125g/4oz piece of pork rind
1 small onion, stuck with 1 clove
1kg/2¼lb lean braising beef
freshly ground black pepper
bouquet garni
salt
pinch of allspice
½ bottle of red wine
300ml/½pt beef stock
25g/1oz beurre manie (12g/½oz butter mashed with 12g/½oz flour)
3tbls tomato purée
24 button onions, glazed, to garnish

- ***Serves 6***

1 In a bowl, combine the finely chopped onions with garlic, shallots and strips of bacon or ham. Grease the base of a heavy 3L/5pt flameproof casserole with butter or chicken fat, and lay the piece of pork rind on top, skin side up.

2 Cut the beef into 3cm/1½in cubes, trimming away any pieces of fat or gristle. Put a layer of cubes on top of the pork rind and season with salt and pepper. Sprinkle with a little of the onion mixture. Continue layering the ingredients until used. Bury the bouquet garni and the whole onion in the centre. Sprinkle with salt, pepper and allspice.

3 Heat the oven to 120C/250F gas ½. Pour the wine and stock into the casserole. Cover and bring slowly to the boil; slip an asbestos mat between the casserole and the source of heat so that it takes about 20 minutes to boil.

4 As soon as tiny bubbles break the surface of the liquid, skim the surface, removing any froth or impurities. Cover the casserole tightly, transfer to the oven for 2½ hours, 160C/325F gas 3.

5 Remove the casserole from the oven and discard the piece of pork rind, the bouquet garni and the whole onion. Stir the *beurre manié* in tiny pieces into the sauce, together with the tomato purée, and simmer gently until the sauce has thickened and no longer tastes of flour.

6 Add the glazed button onions and simmer a few minutes longer. Serve hot from the casserole.

RICH AND FRUITY REDS

PETITE SIRAH

Countries where the sun shines produce wonderful red wines, many of them the result of new high-tech methods

FOR RICHNESS IN wine it is usually the warmer wine-growing regions that shine. Sun-blessed sweet, succulent grapes are more readily turned into fruit-packed wines than their more restrained temperate cousins. But a fair bit of care is needed. If grapes are left to run riot in the sun and over-produce, or if fermentation is allowed to let rip in a too-warm atmosphere, the wines will be dilute and dull.

CHOOSING A CORKSCREW

You would think that in this day and age getting a cork out of a bottle would be simple. But we struggle and tug and still manage to break the cork. The problem is often the corkscrew. There are too many badly designed ones, made for profit or gimmickry rather than as a necessary, hard-working tool.

A good corkscrew depends on two factors. First, it needs a properly made helical screw, shaped like a spiral staircase, that will get a good grip on the cork, not a gimlet-like screw that drives through the middle of the cork, often shredding it to pieces as it is removed, or else refusing to budge. Second, it must give decent leverage so that the implement, not your fingers, does most of the work.

The 'waiters' friend' (top) has a good helix, fair leverage and works well once you are used to handling it. The 'butterfly wing' type (bottom) has great leverage but usually just a cheap gimlet screw. Expensive, but worth every penny, is a patented type with a simple handle (middle) that gets the screw in and the cork out, and turns with ease.

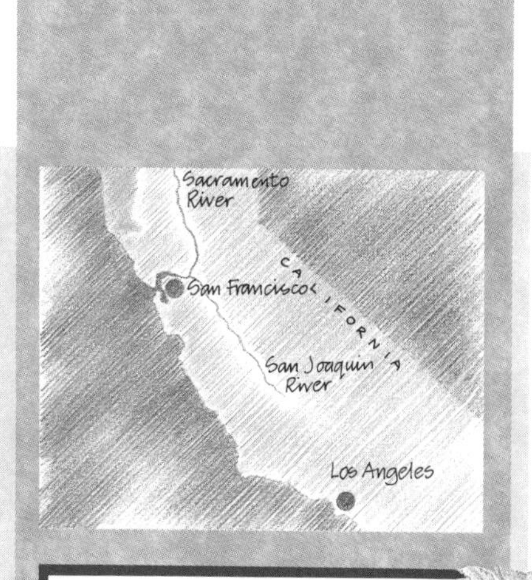

BEST BUYS

Californian Classics Petite Sirah 1980, FBG International Wine Cellars
Taste guide: **full-bodied**
Serve with: **meat pies and casseroles**

Arruda, Adega Cooperativa dos Vinhos
Taste guide: **medium-bodied**
Serve with: **just about anything**

Shiraz-Cabernet 1986, Penfolds Wines
Taste guide: **full-bodied**
Serve with: **pâtés, roast lamb**

Dame Adelaïde 1984
Taste guide: **medium-bodied**
Serve with: **ham, mild cheeses, main-course salads**

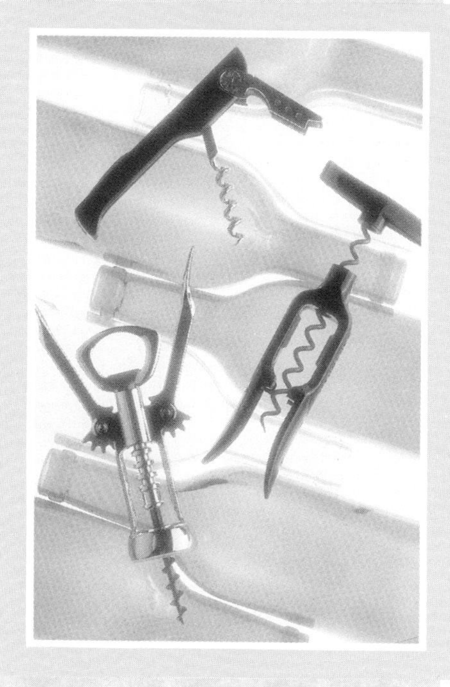

Portuguese innovation

Portugal may have made its name with firm, earthy wines like Dão but it also excels at making succulent, rounded ones, particularly in the southern half of the country where the climate is ideal. A new breed of winemakers, often with experience from abroad, have discovered how to get the most from the native grape varieties and some of the results have astounded even the more cynical wine experts. A terrific example is **Arruda** from the **Adega Cooperativa de Arruda dos Vinhos.** It's a bright, deep ruby colour, and just a sniff brings warmth, ripeness and concentrated berry fruits. A mouthful is not enough. Packed full of flavour, soft but assertive, robust but not too weighty, it demands at least one follow-up mouthful – if not several. What's more, it's excellent value.

Californian classic

California has no difficulty producing fruit in her wines. If anything, the problem is the reverse: preventing an over-abundance of ripe and powerful fruit making the wine too big, fat and jam-like. The problem has been resolved in typical American fashion – by huge investment in winemaking research and technological development. State-of-the-art wineries abound and understanding of production is at a high level. One of the results of this is the **Californian Classics** collection: **Petite Sirah 1980** by FBG International Wine Cellars. A deep red-purple colour, the bouquet bursts out of the glass. Rich is an understatement; it is mouth-filling, intense, with eucalyptus, mint, liquorice and blackcurrants.

Australian original

Half a world away lies Australia, itself no sluggard in getting balanced richness and fruit into its reds. Penfolds Wines is a highly regarded company and their **Shiraz-Cabernet 1986** upholds the reputation. Shiraz and Cabernet are two grape varieties often seen on their own, but blended only in Australia. The partnership works to perfection, the vigorous, weighty and long-lived Shiraz combining with the rounded, velvety, cedary Cabernet to produce a uniquely Australian taste. Shiraz, one of the world's most ancient grape types, is known in France and elsewhere as Syrah but, strangely, it is no relation of the Petite Sirah, the derivation of whose name is a mystery.

Smoky French

But surely France has something to offer? Of course. In a completely different style from the assertive New World wines is **Dame Adelaïde 1984** from the region of Corbières in the south-west. Made from a blend of three varieties well adapted to the area, again including Syrah, it is gentle, smoky, with hints of tobacco. The bouquet is soft, the palate has a little spice and a bit of oak along with the all-important fruit. The wine is made by a special technique to extract plenty of fruit without too much mouth-drying tannin and is then aged for several months in oak barrels.

PERFECT PARTNERS

Yeoman's stew

- ***Preparation:** 45 minutes*
- ***Cooking:** 3 hours*

900g/2lb braising steak, cut into 2.5cm/1in cubes
seasoned flour
4tbls oil
8 pickling onions
2 garlic cloves, finely chopped
8 small carrots, diced
6 celery stalks, sliced
1tsp paprika
1tbls tomato purée
25g/1oz flour
600ml/1pt beef stock
125ml/4fl oz red wine
1 bay leaf
1 bouquet garni
salt and pepper
100g/4oz button mushrooms

- ***Serves 4-6***

1 Coat the beef cubes in seasoned flour. Heat half the oil in a flameproof casserole and fry the beef, in two batches if necessary, turning until brown all over. Remove from the casserole with a slotted spoon and reserve, keeping the meat warm.

2 Add the remaining oil to the casserole and fry the onions, garlic, carrots and celery for 15-20 minutes or until lightly coloured, stirring often.

3 Remove the casserole from the heat and stir in the paprika, tomato purée and flour. Gradually stir in the stock and wine. Add the bay leaf and bouquet garni and season with salt and pepper to taste.

4 Return the meat to the casserole, cover and simmer for 2-2½ hours or until tender, adding the mushrooms after 1½ hours. Serve hot.

CURRY COOLERS

PINOT BLANC

Far from being 'incompatible partners', curry and wine can be matched successfully to complement each other's spices and cool down that burning desire . . .

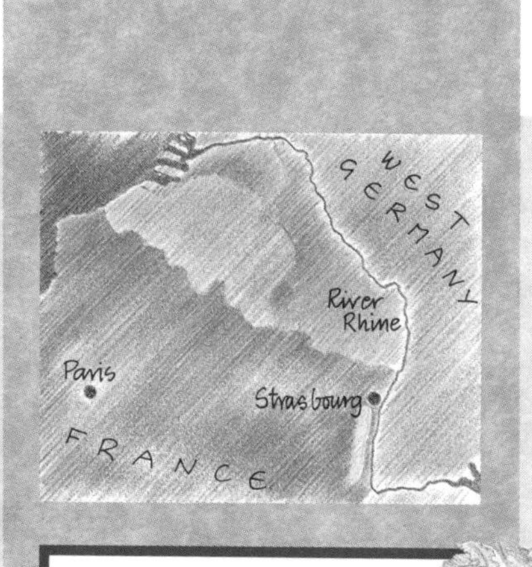

BEST BUYS

Alsace Pinot Blanc
Taste guide: dry

Alsace Gewürztraminer 1987
Taste guide: dry

Les Terres Fines 1987 Cépage Muscat
Taste guide: dry

Côtes de Provence Rosé
Taste guide: dry

IT HAS BEEN said, time and time again, that wine is not a good accompaniment to curry. Lager is the normal recommendation. But is this true or is it just a cop-out by wine smarties who couldn't bear the thought of their esteemed liquid having to do battle with such mouth-dominating fare? And anyway, what if you don't like lager but do want a little alcoholic accompaniment to your favourite Indian dish?

Trying task

A curry wine, though, has to be a pretty clever beast. It has to have enough flavour so that you can still taste it through the curry – otherwise you might just as well drink water – but not so much that it merely adds to the sensations battering your taste buds. It needs to be refreshing enough to help cool you down, but not so refreshing you are tempted to gulp as if it *were* water. It needs nothing, like tannin or dominant acidity, that will make the curry seem hotter, yet has to be well balanced in its own right. It seems an impossible task.

But it isn't. There are certainly wines that do go well with curries and although they won't taste the same as when drunk on their own and may not taste *quite* as good as they normally do, they can certainly make an enjoyable curry accompaniment.

If you like your curry mild, with just a hint of spice, there is really no need to search out special 'curry wines'. The hotter the curry, though, the more wines fall by the wayside. The following all pass the 'heavy on the chilli' test. If, however, your mouth is really burning, a slurp of water first will give the wine more of a sporting chance.

Spice match

The only wine that is usually ever, even tentatively, suggested for curry is Gewürztraminer, most probably because of the theoretical match between spice (from the wine) and spice (from the food). The match does, in fact, work but not with a really strongly spicy Gewürztraminer: instead with the more open, floral, lychee-like, Turkish-delight type such as **Alsace Gewürztraminer 1987** bottled by Les Caves de l'Enfer. Tasted on its own it is full-bodied, but not too weighty, with a well-rounded, dry, gently spicy character: with curry it tastes, if anything, fruitier but is best in small sips.

The revelation, though, and the

best match of all, proved to be also from Alsace, but a different grape variety: **Alsace Pinot Blanc** from the Cave Vinicole Eguisheim. It is quite fat, yet lemony, creamy and a little salty. Perhaps it is the slight saltiness that helps cut the burn, for the wine is a great curry cooler. The impact of its flavour appears less marked with the food, since it has so much taste, and it gains a delicious refreshing character, so much so that it is easy to drink far more of it than you realize at the time.

Stylish partner

Another grape that can rise to the occasion is Muscat, when it is made into a dry wine that isn't too heavy but from a land warm enough for the grape to pick up plenty of ripeness. **Les Terres Fines 1987 Cépage Muscat** from the south of France is such a wine. The deliciously grapey Muscat character is strongly apparent on both bouquet and taste but, rather than being overtly fruity, the fruit is elegant and restrained, giving a firm taste of minerals to the wine. Les Terres Fines is changed remarkably little by its partnering with curry (although it loses a bit of flavour), but has sufficient style to cut clean through the food's tastes.

It is not just white wines that can partner curry. **Côtes de Provence Rosé** works too. As a rosé wine it has practically no tannin – a definite enemy of chilli – but it does have plenty of light, cheering fruit. It is quite deeply coloured for a pink wine – almost light red. The bouquet gives the sensation of warmth, as befits a wine that is produced in the warm Mediterranean provençal part of France, but also a slight herbiness and plenty of rich, strawberry-like fruit. On its own it has plenty of flavour, is ripe, but dry, and without seeing the colour it would be hard to tell if it were a red wine or a white one. Once drunk with curry, the intensity of the flavour lightens, except for the fruit flavours which remain and nicely balance the almost bitter tastes of hot curry spices.

If you really prefer a red wine, or if your curry isn't overly hot and is based on ingredients which would normally be partnered by red wine, the best bet is a 1988 Beaujolais.

For a fish-based curry, such as the coconutty quick fish curry, below, white wines are the best bet. The wine needs to have plenty of flavour though, so a gently spicy, yet fruity Gewürztraminer is a good choice.

PERFECT PARTNERS

Quick fish curry

Most unlikely to feature on any Indian restaurant menu, this unusual light curry is ideal for impromptu home cooking as it is prepared in no time and requires only a few basic curry spices. Serve it with boiled rice and a salad of thinly sliced cucumber

- ***Preparation: defrosting fish, then 15 minutes***
- ***Cooking: 15 minutes***

100g/4oz desiccated coconut
3tbls oil
1 onion, finely chopped
1tsp ground turmeric
1tsp ground ginger
½tsp ground coriander
2tsp flour
2tbs lemon juice
4 frozen fish steaks, defrosted
fresh coriander, to garnish

- ***Serves 4***

1 Put the coconut into a bowl, pour on 300ml/½pt hot water and stir. Leave to soak for 5 minutes, then strain through a sieve into a bowl, pressing the coconut with a spoon to extract all the moisture. Discard the coconut and reserve the liquid.

2 Heat the oil in a frying pan and cook the onion over medium heat for 4-5 minutes, stirring once or twice, until softened. Stir in the spices and flour and cook for a further 2 minutes, stirring constantly.

3 Add the reserved coconut liquid with the lemon juice and bring to the boil, stirring. Add the fish steaks, spooning a little of the sauce over them, and simmer for 10 minutes, uncovered, breaking the fish into large chunks when it is almost cooked.

4 Transfer to a warmed serving dish and garnish with coriander. Serve immediately, on a bed of plain white or pilaff rice, if wished.

WINES FOR CHINESE FOOD

RIESLING HALBTROCKEN

Whether you go for chop suey or deep-fried chilli bean curd, you'll find that many wines make great partners for Chinese food

THERE'S A TENDENCY to think Chinese food is 'difficult' to match with wine. It is probably because traditionally Chinese food is accompanied by other beverages, such as tea, and because Chinese restaurants usually serve little, if any, wine. However, there are many white wines that do make excellent partners for Chinese food. There are just two points of which to beware. If you try to match the sweetness of food (as in sweet-and-sour dishes) with a sweetish wine, either the former will taste sharp or the latter sour, depending on which is the sweeter. And any too powerfully flavoured wine – particularly red – could mask the delicate taste of some dishes.

For wine at its most light and elegant there is no better source than Germany. Grown on some of Europe's most northerly vineyards, the vine struggles to get enough warmth and sun and when the grapes ripen (which some years they don't) the result is unparalleled fruity acidity and finesse. Most German wines are medium dry or medium sweet and can be sweet or, occasionally, very sweet. Recently drier wines have become more popular. The driest are called Trocken and the dryish Halbtrocken and these are the ones that are best for drinking with a meal, particularly one that is Chinese.

BEST BUYS

Riesling Hock Rheinpfalz Qualitätswein Halbtrocken
Taste guide: *dry*

Baden
Taste guide: *dry*

Alsace Gewürztraminer 1986
Taste guide: *dry*

Babich 1987 New Zealand Sauvignon Blanc Hawke's Bay
Taste guide: *very dry*

Reasoning with Riesling

Riesling Hock Rheinpfalz Qualitätswein Halbtrocken by Scholl & Hillebrand in their Corkscrew Series is an ideal German wine for Chinese food. It is from the Riesling grape, Germany's best and much in demand, and has the characteristic racy, lively aroma of the grape that is both fruity and oily (many people say it reminds them of petrol – and yet they still love it!). Hock is a name for any of the German wines from the Rhineland areas, in this case the Rheinpfalz. It is not the coolest of the German wine areas, being protected by the Haardt mountains, and so is well suited to producing wines on the drier side; sweetness is usually essential to balance the very high acidity of grapes grown on the coldest sites. The flavour is neither too powerful nor too weak. It is fresh, a mouthful of crisp fruitiness that cleans the mouth and cuts any fat in the dish. A hint of its taste remains, leaving the drinker ready for his or her next bite of food.

Fruit and flowers

Further south in Germany lies Baden, the most southern of the country's vineyard areas. The greater amounts of sun and warmth Baden receives means it produces fuller, meatier wines – yet still very much in the German mould. The wine is labelled simply **Baden** and comes predominantly from the Müller-Thurgau grape variety. Flowery, a bit like honeysuckle, and aromatic to smell, the taste is dryish, quite firm with clean apple and greengage fruitiness. A good match for most Chinese dishes, especially those involving finely sliced meats, it will enhance their flavours without intruding.

The French connection

Just across the Rhine river to the west from Baden is Alsace in France. It is one of the sunniest and driest areas of France because it lies to the east of the Vosges mountains which form a protective rain shadow. Alsace produces wines with all the aroma of Germany but the structure of France – a winning combination but, surprisingly, one that has been slow to catch on. They are dry, sometimes quite powerful, but always clean and pure-tasting. Many of the grape varieties grown are similar to those across the river in Germany but one, Gewürztraminer, has almost become known as Alsace's own. **Alsace Gewürztraminer 1986** from Cave Cooperative d'Ingersheim is typical, with an inviting bouquet – just like lychees with spice – which is carried through to the taste. The natural affinity of these flavours with Chinese dishes is obvious, but Gewürztraminer can be full-bodied, strong in alcohol and sometimes almost overbearingly spicy. This wine, however, is lighter in style, with the elegance to partner, not overpower, the food. Although dry, its richness makes it appear a little sweet.

Wine from down under

For a real treat, try a wine from the other side of the world. **Babich 1987 New Zealand Sauvignon Blanc Hawke's Bay** is one of those extremely moreish wines that it is difficult to stop drinking. The Sauvignon grape variety originally comes from northern France where its characteristic gooseberry flavour is epitomized by Sancerre. Thousands of miles away, in Hawke's Bay in New Zealand, the climate is suitably cool for the grape - but, overall, it is just a touch warmer and somewhat sunnier than northern France. So the wine keeps the gooseberry flavours, but also gains ripeness.

PERFECT PARTNERS

Sliced beef with oyster sauce

Serve this colourful stir-fry as part of a Chinese meal with boiled rice and a fish or vegetable dish.

- ***Preparation: 35 minutes, plus 20 minutes marinating***
- ***Cooking: 5 minutes***

225g/8oz rump steak
2tbls oyster sauce
1tbls dry sherry
1tbls cornflour
4tbls oil
2.5cm/1in piece of fresh root ginger, finely chopped
1 garlic clove, finely chopped
1 large onion, halved, sliced and separated into rings
1 red pepper, seeded and thinly sliced
1 green pepper, seeded and thinly sliced
1tsp salt
1tsp sugar
2tbls chicken stock or water, if necessary
2 spring onion tassels, to garnish
boiled rice, to serve

- ***Serves 4*** 

1 Cut the beef into thin slices, each measuring about 5cm × 2.5cm × 3mm/2in × 1in × ⅛in.

2 In a shallow dish, mix together the oyster sauce, sherry and cornflour and add the beef slices. Stir to coat thoroughly, then leave the beef to marinate for about 20 minutes.

3 Heat half the oil in a wok or a large, heavy frying pan. When it is very hot, add the beef and stir-fry for 10-15 seconds, until it is just cooked through. Remove with a slotted spoon and reserve.

4 Add the remaining oil to the wok or pan, heat, then add the ginger and garlic and stir-fry for 1 minute. Add the onion, red and green peppers, salt and sugar and stir-fry for 1½-2 minutes or until softened but still crisp.

5 Add the beef slices and blend well together, adding a little stock or water if necessary. Serve the beef hot, garnished with spring onion tassels and accompanied by boiled rice.

AUSTRALIAN RIESLING

It is always a good idea to keep a selection of wines for any occasion: here are some whites to suit all palates

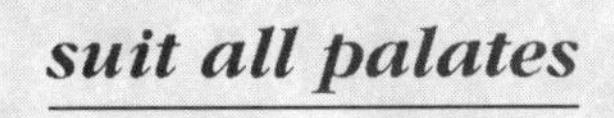

BEST BUYS

Australian Padthaway Rhine Riesling
***Taste guide:* dry**

Muscadet de Sèvre et Maine 1988 Domaine de la Grange
***Taste guide:* very dry**

Vermentino di Sardegna
***Taste guide:* dry**

Château de la Tuilerie 1988 Entre-Deux-Mers
***Taste guide:* very dry**

Château Roquetaillade La Grange 1987 Graves
***Taste guide:* dry**

ALTHOUGH IT IS more likely that if you have decided to keep some wine at home, to be ready for any occasion, you will have stocked up on red, there is absolutely no reason why you should not keep some white too. Red is the wine most people think of when they talk about 'laying down' wine, simply because much more of it matures a lot slower and lasts a lot longer than most whites. But as long as you buy no more than enough to suit your needs over the next six months or so you can enjoy all the benefits of wine at home and never need to dash out to the off-licence.

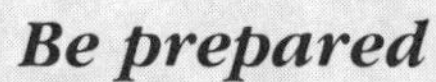

Be prepared

As long as you have a bit of space, the ideal is to keep one or two bottles of white wine in the fridge, to be ready chilled. The wine will, in fact, be a little too cold but it will warm up very quickly in a warm glass.

The most versatile type of white is one which is light, so not too demanding, and crisp, so refreshing. There is no better example of this than a good Muscadet like **Muscadet de Sèvre et Maine 1988 Domaine de la Grange.** Muscadet is the name of the grape variety and also of the area, towards the mouth of the Loire river in north west France, around the town of Nantes. The Sèvre and the Maine are two tributaries of the Loire and the lands drained by them produce the best Muscadet. As the area is well known for its oysters Muscadet does well with them. But, in fact, it is great with all sorts of fish and sea food as well as being one of the best wines for drinking on its own. A zingy, appley fresh bouquet and an appley, peachy, greengage taste make this the sort of wine you will be happy to drink on its own any time.

Similarly fresh, but even lighter in weight is **Vermentino di Sardegna.** Sardinian wines traditionally were big, heavy, alcoholic and often sweet. Now, with newly acquired skills and helped by the technology that allows them to control all aspects of the wine-making, Sardinia's winemakers have discovered that their local grapes make delicious, crisp, easy-drinking everyday wines. As its origins suggest, Vermentino is ideal for warm weather – or for warm rooms when it is cool outside. Lemony-appley, crisp but with a well rounded finish and a

pleasingly bitter finish, the wine is just the thing for a quiet drink with friends, with snacks or light meals or on its own. You won't tire of it.

Entre-Deux-Mers

For a little more weight, without losing any enlivening acidity, and a wine that will go with a broad range of foods, try **Château la Tuilerie 1988 Entre-Deux-Mers.** Entre-Deux-Mers means 'between two seas' but the land where the grapes are grown actually lies between two rivers, the Garonne and Dordogne in south west France's Bordeaux region. As the rivers are tidal and join the Gironde which flows into the Atlantic ocean, perhaps the term 'seas' is not too much of an exaggeration. All Entre-Deux-Mers wines are dry. They are made from the Sauvignon and Sémillon grape varieties. The Sauvignon, in particular, gives a silky feel to the wine and a penetrating taste reminiscent of gooseberries, while the Sémillon softens down the pungency with a lanolin creaminess. Château la Tuilerie is a very good, very more-ish example. As a stand-by it can be pressed into service on numerous occasions and will always go down well with various dishes and those unexpected guests.

Dry or sweet

What do you do when some of your friends like dry wines and some of them prefer sweet? What wine could suit both groups? **Australian Padthaway Rhine Riesling** is the answer. Although the wine is dry – dry enough for anyone who prefers never to drink anything sweet – it has such a round, ripe fruitiness that it appears sweeter than it is. It will even suit those who reckon most dry white tastes sour and who always plump for quite a sweet taste. It is made from the same Riesling grape variety that is responsible for Germany's finest wines. Grown, though, in the warmer climate of the southern part of Australia, in Victoria, near Melbourne, the grapes get much fuller and riper and pick up a minerally, tropical fruits flavour which is quite captivating.

Impromptu entertaining

If one of your reasons for having a few bottles of white wine ready at home is always to be prepared for impromptu dinner parties, whether at your home or with others, you may consider spending just a little bit more on it. **Château Roquetaillade La Grange 1987 Graves** is then the one to choose. It comes from Bordeaux but the Graves district which produces Bordeaux's best whites. It is made from the Sémillon grape variety, aged for a short time in new oak barrels to give it extra fleshiness, roundness and class. It has a fruit-packed but gently toasty bouquet and a harmonious, crisp but mellow taste in which you discover more flavours the more you try it.

If you have unexpected guests arrive, there's really no need to resort to a take-away supper. Cook a quick supper dish such as Indonesian skewers, below. Defrost the fish in the microwave if you are really short of time, and reduce the marinating time from 30 to 10 minutes, if necessary.

PERFECT PARTNERS

Indonesian skewers

- ***Preparation: 20 minutes plus 30 minutes marinating***
- ***Cooking: 10 minutes***

4 frozen fish steaks, defrosted
3 courgettes, cut into 2.5cm/1in slices

For the marinade:
50g/2oz desiccated coconut
1 garlic clove, crushed
½tsp ground ginger
½tsp light soft brown sugar
½tsp coriander seeds, crushed
2tsp soy sauce
2tsp lemon juice

For the garnish:
2 lemon slices, quartered
2 fresh parsley sprigs

- ***Serves 4***

1 Cut the fish into 4cm/1½in cubes. Thread the cubes and courgette slices onto 4 wooden skewers and place them in a shallow dish.

2 Make the marinade: put the coconut into a bowl, pour on 7-8tbls hot water and stir. Leave to soak for 5 minutes, then strain through a sieve into a bowl or jug, pressing the coconut with a spoon to extract all the liquid. Discard the coconut and add the garlic, ginger, brown sugar, coriander seeds, soy sauce and lemon juice to the coconut liquid.

3 Pour the marinade over the skewers and turn them to coat thoroughly. Refrigerate for 30 minutes.

4 Heat the grill to medium. Remove the fish skewers from the marinade and grill for 7-8 minutes, until cooked, turning frequently and brush with the marinade.

5 Arrange the skewers and serve at once on heated plates.

BUBBLE LOVE

MONISTROL ROSE

Bubbly wine creates the feeling of celebration and pink bubbles make everyone feel extra special

BEST BUYS

Marqués de Monistrol Rosé
***Taste guide:* dry**

Champagne Gall Brut Rosé
***Taste guide:* very dry**

Angas Brut (Australian)
***Taste guide:* dry**

Tocade Framboise
***Taste guide:* medium dry**

Domaine de Fourn Blanquette de Limoux
***Taste guide:* dry**

Asti Spumante Arione
***Taste guide:* sweet**

W*HEN THE REASON* for cracking a bottle has something to do with love: a new lover, an anniversary, a way to say 'I'm sorry', or simply just a reminder to someone how important he or she is to you, you will get the message across best if the wine is sparkling, and probably pink too. There is no particular reason why bubbles should make a wine so much more right for lovers, but they do – they create the feeling that the drinkers are special and indulged and they induce a touch of celebration into the proceedings. As for the pink colour, it may be just the prettiness of the hue or it may be that rosé sparkling wines seem more select – the reasons for the impression don't matter as long as it is the right one.

Wines for lovers

Sparkling wine is usually quite expensive. It has to be because the better methods of making it are time-consuming as well as being skilled and labour-intensive jobs. Once the price of a bottle heads over the £5 barrier (and many do) it is tempting to go the whole hog and splash out on a bottle of Champagne. After all, once in a while, for someone very special, why not be a little extravagant? You can't fail to enjoy **Champagne de Saint Gall Brut Rosé** (Brut means that it is dry). It is a delicate coral colour, with plenty of fizz, but tiny, tiny bubbles that tickle your palate rather than bombard it with an explosion of gas. When first opened it has very little bouquet but, as it sits in the glass, more and more biscuity, yeasty, strawberry, lime aroma becomes apparent. Dry, with a rich but restrained taste and the same flavours you can smell, it is a remarkably refined wine.

Coming back down to earth **Marqués de Monistrol Cava Rosé Brut** should get a few eyes glistening. Cava is the term used in Spain to indicate that a sparkling wine has been made by the best method, as in Champagne, which may be called méthode champenoise, méthode traditionelle or método clássico elsewhere. Marqués de Monistrol, comes from north east Spain, in Catalonia, not far from Barcelona, where nearly all the best Spanish sparkling wines are produced. A mid-depth rose pink, it has a cheeringly fruity, strawberry and raspberry nose with a rounded, fruit-packed, dry and quite restrained flavour and plenty of weight. Its bubbles are long-lasting and not too aggressive, giving plenty of fizz in the mouth but not so much that it is hard to taste the wine. Overall, it is one of the best balanced and most attractive rosé sparklers around.

For something particularly unusual, another rosé should make you both sit up and take notice. It's **Angas Brut** from Yalumba and comes from the Barossa valley of Australia. Australian wines are now rightly popular as their emphasis is on ripe, delicious fruitiness from beginning to end. A deep salmon colour, the aromas of citrus and tropical fruits waft by as soon as the glass is lifted to the nose. Its taste is quite full, rich and distinctively fruity, so ripe it seems a touch sweet but its finish is cleanly dry. It's just the wine if you want to share an evening full of laughter rather than meaningful conversation.

Sparkling eyes

If you reckon you are going to be too busy staring into each other's eyes to pay much attention to what you are drinking, choose something where you can get lots of enjoyment just from sniffing – you can stare and sniff at the same time! **Tocade Framboise** might just be the right idea. It is a drink made from a sparkling wine base enlivened by added raspberry aromas. The aroma of raspberry is quite strong so, as long as you like raspberry, you will get plenty of pleasure from a glassful. It tastes a little like raspberry as well, with some lime too and a bitter-sweet tang.

Of course, not everyone goes for rosé wines. If you or your partner don't, you could create just the right atmosphere with **Domaine de Fourn Blanquette de Limoux.** Blanquette de Limoux comes from the south of France and can be one of France's best sparkling wines. When just bottled it is often very appley but, with a little time it turns into a wine like Domaine de Fourn: a small, even and persistent stream of bubbles; a nutty, yeasty, lightly grassy bouquet and a rounded, grassy-biscuity flavour with just a hint of apples remaining.

Sweet and bubbly

What about something as delightful as a sparkling wine reminiscent of lightly whisked-up muscat grapes to accompany your time together? **Asti Spumanti Arione** fits the bill perfectly. Asti Spumanti comes from around the town of Asti in north-west Italy where the Moscato grape variety produces some of the most delicious, light, sweet, delicately aromatic, musky, grape wines it is possible to find. Fresh as a daisy, floral, clean and pure-tasting and low in alcohol too, you could drink more of this wine and stay clear-headed. Asti Spumanti Arione is not just a wine for lovers, it is a wine to turn friends into lovers.

To create the right mood, the food needs to be romantic too. Caviar is the perfect partner to bubbly wines and is classically served with Russian blinis (see below). These little pancakes take a while to prepare as the batter needs to be left to rise, but they can be prepared well in advance.

PERFECT PARTNERS

Russian blinis

- ***Preparation: 3 hours, including rising time***
- ***Cooking: 9 minutes***

275ml/10fl oz milk
1½tsp dried yeast
2tsp sugar
50g/2oz flour, sifted
50g/2oz buckwheat flour, sifted
¼tsp salt
2 large eggs, separated
15g/½oz melted butter and butter for greasing
For serving
lemon wedges
soured cream
red caviar, smoked sturgeon or smoked salmon

- ***Makes 30***

1 Heat 150ml/5fl oz milk until lukewarm and combine 50ml/2fl oz of the warmed milk in a warm bowl with the yeast and ½tsp of the sugar. Leave to stand for 10 minutes until the yeast foams. Sift together the flours and the salt. Add the yeast liquid and the remaining 75ml/3fl oz of warmed milk. Mix well. Cover the bowl, stand in a warm place and let the batter rise for about 2½ hours.

2 Warm the remaining milk to just lukewarm and whisk in the egg yolks. Add the melted butter and the remaining sugar and add to the batter mixture. Beat well and let the batter stand, covered, for 30 minutes in a warm place.

3 When ready to cook the batter, whisk the egg whites until stiff but not dry and fold them into the batter. Heat a griddle, or an iron Swedish *platter* pan with indentations for cakes, and butter it lightly. Alternatively, use a heavy frying pan with buttered 5cm/2in biscuit cutters to shape the blinis.

4 Spoon 1tbls batter at a time on to the hot griddle and cook for about 3 minutes. The blinis should not be more than 5cm/2in in diameter. Flip them over with a fish slice and cook for 3 minutes on the second side. Transfer the blinis to a warmed platter as they are cooked.

5 Butter the griddle again lightly, as necessary, but take care not to use too much. Serve the blinis hot with lemon wedges, soured cream and red caviar, smoked sturgeon or smoked salmon.

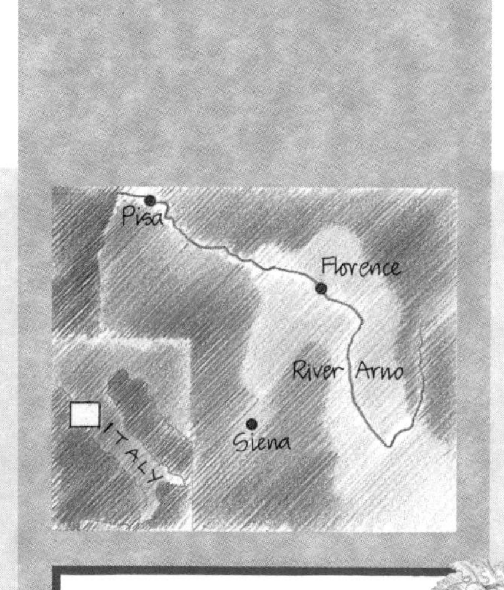

CHIANTI RUFINA 1986

Down-to-earth red wines meet the demands of pizza best of all, especially when they come from the Chianti

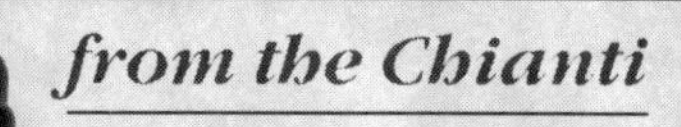

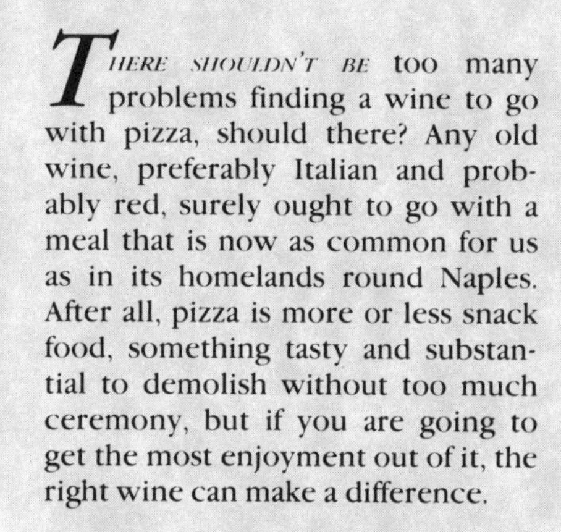

BEST BUYS

Fattoria Selvapiana Chianti Rufina 1986 (🍷🍷)
***Taste guide:* medium-bodied**

Rocca delle Macie Chianti Classico 1987 (🍷🍷)
***Taste guide:* medium-bodied**

Côtes du Roussillon Les Quatre Coteaux (🍷)
***Taste guide:* light-bodied**

Pedregal Rioja 1988 (🍷🍷)
***Taste guide:* medium-bodied**

Frascati Superiore Estate bottled, Azienda Vinicola SAITA (🍷🍷)
***Taste guide:* dry**

THERE SHOULDN'T BE too many problems finding a wine to go with pizza, should there? Any old wine, preferably Italian and probably red, surely ought to go with a meal that is now as common for us as in its homelands round Naples. After all, pizza is more or less snack food, something tasty and substantial to demolish without too much ceremony, but if you are going to get the most enjoyment out of it, the right wine can make a difference.

Chianti, prego!

A simple pizza margherita (tomato and cheese) will be amenable to most wines, but the more topping you include, the more chance a wine won't go too well. The toughest test is against anchovy and as anchovy is one of the best ingredients to add a bit of zip to a pizza, it is as well to choose a wine that can cope. The best wines are red, youngish, punchy but not too full-bodied and with good acidity. The obvious choice seems to be Chianti – but which Chianti? With many dishes, it would be best to go for a Chianti Riserva which is aged longer than normal in large oak barrels and has a more mellow style, but pizza is better with something younger and livelier. A Classico could fill the bill. This is a wine from the heartland of the Chianti district between Siena and Florence. You won't go far wrong with **Rocca delle Macie Chianti Classico 1987.** It is just the sort of attractive zippy wine that you could enjoy with a variety of dishes: it cuts through any richness of the pizza topping and balances the saltiness of any anchovies. The wine is made predominantly from the Sangiovese grape which gives it a violetty, cinnamon bouquet and a rich, spicily fruity, plummy taste with enlivening acidity.

Of the other parts of the Chianti district surrounding the Classico, just one, called Rufina (not to be confused with an estate with a very similar name) can produce wines every bit as good, if not better, than those from Chianti Classico. Chianti Rufina lies east of Florence with vineyards on high slopes that lead down to the valley of the Sieve river. Often, Rufina wines have greater firmness and acidity than those in Classico (a good sign for pizza wines) and longer ageing potential too. So **Fattoria Selvapiana Chianti Rufina 1986** could hardly be bettered. One year older than the Rocca delle Macie, it has a little more fleshiness to balance its acidity. With a bouquet that includes plums, spice, cinnamon and tea, and a firm, rounded taste that combines well with the tomatoes and the pizza base and contrasts well with the anchovies, having just the right weight to balance them, this wine will turn a pizza into a feast.

French foils

Of course, it is not just the Italian wines that can be great with pizza. From France, try **Côtes du Roussillon Les Quatre Coteaux.** Côtes du Roussillon is in the far south west of France, near the Mediterranean coast. The grapes for this wine are grown on the foothills of the Pyrenees and come from four different hill communes from which they are carefully selected. The wine has an attractive minerally, creamy blackberry bouquet and a lightish, dusty, peppery fruit taste that makes it a good, gentle foil for the assertive pizza flavours. When faced with the challenge of anchovy, it tastes softer, rounder and better and it helps cut the anchovy's saltiness too.

Spanish partners

Spain, too, can provide a good pizza partner: Rioja. But, as with Chianti, it is not the mellow, well-aged Riojas that match best. Instead, a young wine, like **Pedregal Rioja 1988**, which has not developed the rich vanillic oakiness that typifies older Riojas, is the one to go for. Pedregal has a rich, alluring, nuts, limes and plums bouquet. It has a firm, plumskins, blackcurrant and cooked apples flavour with a touch of leanness which disappears as soon as it is drunk with the varied flavours of a pizza. It is not too powerful for the simplest, lightest pizza, it passes the anchovy test with aplomb, and it makes the pizza flavours seem even more focused.

A touch of white

What about a white with pizza? There is no reason why not, even if red is the most common accompaniment. Working by the almost infallible rule that dishes traditional to certain areas are matched best by wines from the same areas, try Frascati. The area around Naples, pizza's homeland, is the Italian region of Campania, but few white wines are exported from there. Frascati, though, comes from just a few miles further north, from the hills south east of Rome. **Frascati superiore** estate bottled by Azienda Vinicola SAITA is just right. Cool and refreshing, its crisp but soft flavours produce a stimulating contrast to the richness of the pizza and it softens the attack of anchovy without damaging the flavour.

If you haven't much time to make a pizza, then use a quick scone base in place of the bread dough: mix 225g/8oz self-raising flour, ½tsp salt and 2tbsp oil to a soft dough with about 150ml/¼pt milk. Turn out on to a floured surface and knead lightly. Roll out to a 30cm/12in round and then continue from step 3, below.

PERFECT PARTNERS

Neapolitan pizza

- ***Preparation:** 1¼ hours, including rising-time*
- ***Cooking:** 20 minutes*

1tsp easy-blend dried yeast
½tsp sugar
250g/8oz flour
1tsp salt
1tbls olive oil
150ml/¼pt milk and water

For the filling:
400g/14oz canned tomatoes
1tsp dried oregano
salt
freshly ground black pepper
175g/6oz Mozzarella cheese, sliced
50g/2oz canned anchovy fillets, drained and cut into strips
about 15 black olives

- ***Serves 4*** 

1 To make the pizza base, work the ingredients together and mix with enough just-warm milk and water to form a pliable dough, then turn out on to a floured surface. Knead for 5-10 minutes until the dough is smooth and elastic. Transfer to a large clean bowl, cover with a cloth and leave in a warm place for about 1 hour, until the dough has doubled in size.

2 Heat the oven to 220C/425F/gas 7 and oil a baking sheet. Knock back the dough, then roll out to a 30cm/12in round, 6mm/¼in thick. Place on the sheet and brush with oil.

3 Drain the tomatoes, break them up and spoon on to the pizza base. Season with oregano and salt and pepper to taste. Cover the tomatoes with the Mozzarella slices, then top with the anchovy strips, arranged in a lattice pattern, and the olives. Spoon over the remaining oil and bake for 20 minutes.

WINES FOR PASTA

CHIANTI CLASSICO

Garlic and tomatoes are the basic ingredients for most pasta sauces, but not all wines respond well to these flavours

PRACTICALLY ANY WINE you care to name will go with pasta itself, for pasta is as gently flavoured as potato and acts as a base for anything else eaten or drunk with it. It is the *sauce* that matters. A fishy sauce, for example, will need wines that normally go with fish, a meaty one needs wines at home with meat dishes. Yet when pasta is mentioned it usually conjures up the image of a typically Italian pasta dish whose sauce, whatever else it may contain, is based on tomatoes, well seasoned with garlic, and covered with Parmesan cheese.

The simple touch

It is these deceptively simple ingredients which determine what is, and isn't, a successful pasta wine, and they can wreak more havoc on the taste of some wines than you would ever imagine. Tomatoes, once concentrated – either by cooking or as tomato purée – are quite powerfully flavoured and can be pretty sharp, cancelling out the softness in some wines. Garlic makes a strong impression on the taste buds, and stays there, and can far too easily make a wine seem metallic.

As a general rule, ripe fruity wines from the Cabernet grape are best kept for less rustic fare as they don't respond well to the garlic/tomato treatment. Thankfully, there are plenty of other wines that rise to the challenge magnificently. As there so often seems to be an inborn marriage between wines and foods originating in the same part of the world, Italian wine would seem a good bet – and it is.

Chianti classics

The first name to come to many people's minds when thinking about something to drink with pasta is Chianti – and the instinct is absolutely right. There is a natural affinity between the firm, somewhat acidic, assertive wine that is Chianti, with its overtones of tea and cinnamon, and the sharp, simple flavours of a basic pasta sauce. Few Chiantis fail the pasta test. To see the combination at its best there's no need to buy one of the more expensive Chiantis, marked *Riserva*. (Riserva means the wine is of particularly good quality and has been matured for longer than normal.) Just go for a

ITALY
Florence
River Arno
Rome

BEST BUYS

Chianti Classico 1986
Taste guide: light-bodied

Vigneti di Marano Valpolicella 1985
Taste guide: medium bodied

Montepulciano d'Abruzzo 1987
Taste guide: full bodied

Esteva 1985
Taste guide: full bodied

Côtes du Roussillon 1986
Taste guide: medium bodied

normal Chianti, preferably a Chianti Classico (from the central classic heartland, or Classico, area of the Chianti district in Tuscany, between Florence and Siena). An ideal example of pasta-loving Chianti is **Chianti Classico 1986**, bottled in Gaiole.

Superior Valpolicella

If Chianti is the most recognizable Italian red wine name, Valpolicella must be the runner-up. Valpolicella, too, can be a great pasta wine, but it is better to go for one of the more mellow, more concentratedly flavoured Valpolicellas like **Vigneti di Marano Valpolicella Classico Superiore 1985** from Boscaini. Classico refers, just as for Chianti, to the central heartland zone, this time in north east Italy, north of the Romeo-and-Juliet town of Verona. Superiore means the wine has just a little more alcohol than the norm, and has been aged at least a year before being sold so all its flavour components are developed and well harmonized. The important thing about this wine, though, is that it comes from the special vineyards called Marano, and from carefully tended grapes. Deeply coloured, with a gentle maraschino cherry-like bouquet and a rounded, bitter-sweet taste, it will make your pasta taste terrific.

One of Italy's most exciting, most fruit-filled, 'energetic' grapes is Montepulciano. When grown along the Adriatic coast, in the central south region called Abruzzo, it makes incredibly moreish, deeply coloured, cheering wines, which are flavour-packed, full-bodied and hearty without being heavy. It's a great pasta wine too, with enough of its own zip to combat the punchiest *sugo*. For the essence of Montepulciano, try **Montepulciano d'Abruzzo 1987** from Cantina Tollo. It has an intense purple-red hue, a nose like redcurrants with celery, and a round, vibrant, mouth-filling taste of berried fruits that gives as much of a lift to your mood as your spaghetti. Excellent value too.

Yet though Italy produces some of the best pasta wines, it hasn't completely cornered the market. Numerous Portuguese reds, like **Esteva 1985** from the Douro region produced by Ferreira, go down a treat with the robust flavours of the staple of the Mediterranean diet. Douro is the part of Portugal, named after the river of the same name, where port comes from – but it also makes plenty of other good red wines.

The gentle touch

Neither should France be ignored, particularly if your preference is for a softer, lighter wine when you eat. The wines from the south and extreme south west are the best, as they have all the flavour that the warm sun gives them and come from pasta-friendly grape varieties. **Côtes du Roussillon 1986 Jean-Paul Bartier** is an excellent, and good value, example. Mid-depth ruby, with a gentle smell of hedgerows and herbs, makes a gentle impression in the mouth too, yet its taste is persistent. It is a perfect foil for all sorts of strongly flavoured pasta sauces from simple olive oil and cheese mixtures to the rich and meaty sauce, below.

PERFECT PARTNERS

Spaghetti with rich meat sauce

- ***Preparation: 20 minutes***
- ***Cooking: 3¼ hours***

2tbls olive oil
100g/4oz pancetta or unsmoked streaky bacon, very finely chopped
1 garlic clove, finely chopped
1 small onion, finely chopped
1 small carrot, finely chopped
1 small celery stalk, finely chopped
200g/7oz lean minced beef
150g/5oz lean minced pork
75ml/3fl oz red wine
1½tbls tomato purée
200ml/7fl oz beef stock
salt and pepper
boiled spaghetti, to serve
grated Parmesan cheese, to sprinkle

- ***Serves 4***

1 Heat the oil in a deep, heavy saucepan and fry the pancetta or bacon gently for 2-3 minutes. Add the garlic and vegetables and continue frying over medium heat until they are soft, stirring frequently.

2 Add the minced meats and cook for 5 minutes or until they have lost their pinkness, stirring constantly. Pour in the wine and boil briskly for 2-3 minutes.

3 Add the tomato purée and stock, season to taste, then bring to the boil. Stir thoroughly then simmer, covered, over very low heat for 2-3 hours, stirring occasionally. Serve over boiled spaghetti, sprinkled with grated Parmesan.

CHEAP AND CHEERFUL REDS

COTEAUX DU QUERCY

No need to go for broke when you buy a good red wine: there are great bargains from the Vins de Pays

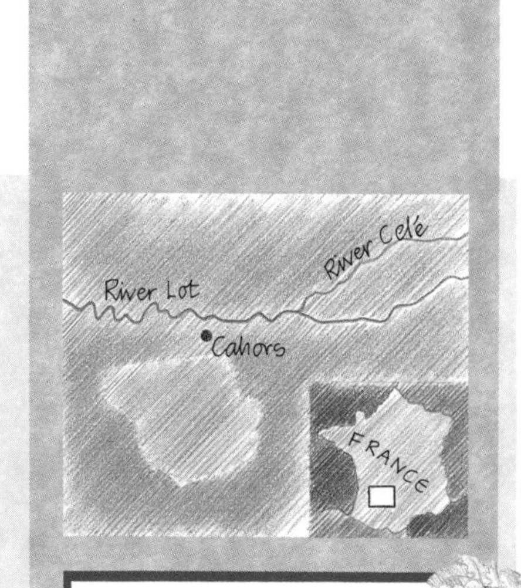

BEST BUYS

Vin de Pays des Coteaux du Quercy, Rigal et Fils
***Taste guide:* medium-bodied**

Vin de Pays de L'Aude
***Taste guide:* light-bodied**

Montevino 1988 Merlot del Veneto
***Taste guide:* light-bodied**

Hungarian Merlot
***Taste guide:* medium-bodied**

Don Hugo
***Taste guide:* medium-bodied**

YOU'RE MISERABLE AND you want to do something about it. You fancy some wine, a drop of something with a good taste that will help you forget your woes. But you're also rather broke – so can you afford it? Yes – if you go for red. While it is not easy to find first-class white wine without breaking the £3 barrier, you can find plenty of wonderfully cheery fruity reds at a great deal less than that.

A great mouthful

The name **Vin de Pays des Coteaux du Quercy, Rigal et Fils** is quite a mouthful, but so is the wine. Both are well worth getting your tongue round. The Coteaux du Quercy, in the interior of south west France, is one of the many Vin de Pays regions that cover most of France's vineyard areas. The wines are nearly always good value.

Rigal is one of the area's best-known and best-regarded producers. The wine is a deep purply red with a terrific creamy-raspberry brackeny bouquet that is an incitement to immediate slurping. The taste is dry but fruit-packed, firm but fleshy and with a touch of cinnamon; this is the kind of wine you could get hooked on. Drink it on its own or try it with red meat – casseroles, meat sauces or sausages.

Pinky-red, strawberry-soft

Another Vin de Pays that says 'begone, dull care' is **Vin de Pays de L'Aude**, from the Union des Grands Chais. The Aude, a *département* in France's far south, where the Mediterranean meets the Pyrenees, produces much red wine that can be ordinary or splendid, depending on the producer. Vin de Pays de l'Aude is bright pinky-red, with a stony, mineral, hot-earth, strawberry bouquet and a softly rounded flavour of strawberries and spice. Drink this wine with chicken, pork or meat pies.

Fruit-cake aroma

The Merlot is a most useful red grape. Grown with great care, it can produce magnificent wines – at magnificent prices! But it can also give us some very affordable light charmers. Italy is a major source of Merlot wine and **Montevino 1988 Merlot del Veneto** is one such. The Veneto region runs from Venice in the east, through Padua and Verona, to Lake Garda in the west.

Like many Merlot wines the Montevino Merlot has a fruit-cake aroma, but it also has a savoury tang, a hint of pepper and a warm ripeness. The taste is dry, somewhat meaty and with the slight bitterness of many north east Italian wines. But its gentle, sweetly-soft fruit compensates harmoniously for any austerity. Like many Italian wines it is meant to be drunk with food – try it with pasta – but it's splendid on its own.

Fresh and lively

Hungary too makes a Merlot wine, called simply **Hungarian Merlot.** It is from the country's extreme southerly region, Villany, which has a 'Mediterranean' climate well suited to the grape. This wine is so richly fruity that one sniff will convince you its been made from blackberries. The fruit is also up front in flavour. The wine is fresh, herby, lively and very good value.

Oaky opulence

There is something very satisfying about the vanilla oaky opulence of wines matured in oak barrels. Unhappily oak-matured wines are usually more expensive than others and the newer the oak, the more toasty and voluptuous the wine.

You won't get new oak flavours from any of the cheap and cheerful wines, but you can get more than a hint of it at a low price from Spain. **Don Hugo** is made from the Tempranillo grape, whose natural flavour seems to enhance any oakiness. The wine is medium to light red with a ripe vanilla bouquet and a softly rounded, sweet fruit taste.

What better to serve with one of these cheap and cheerful reds than fettucine alla bolognese (see below). The recipe contains wine too, and it is always a good idea to use the same red that you will be drinking so that the flavours complement each other.

PERFECT PARTNERS

Fettucine alla bolognese

- ***Preparation: 15 minutes***
- ***Cooking: 40 minutes***

450g/1lb fettucine, dried or fresh
salt
freshly grated Parmesan cheese
For the bolognese sauce:
25g/1oz butter
1tbls olive oil
1 onion, finely chopped
225g/8oz lean minced beef
salt and freshly ground black pepper
1/4tsp dried oregano
150ml/1/4pt red wine
3tbls tomato purée
1tbls chopped parsley

- ***Serves 4***

1. To make the sauce heat the butter and olive oil in a large frying pan. Add the finely chopped onion and sauté over medium heat for 7-10 minutes or until soft, stirring occasionally with a wooden spoon.
2. Stir in the lean minced beef and sauté for 5 minutes or until browned, stirring constantly. Season with salt and freshly ground black pepper to taste, and the dried oregano.
3. Stir in the red wine and tomato purée and simmer gently, uncovered, for 35 minutes. Sprinkle with parsley.
4. Meanwhile, bring a large saucepan of salted water to the boil, add the fettucine and cook for 4-10 minutes depending on whether it is dried or fresh, or until al dente, cooked but still firm.
5. Drain the pasta thoroughly and transfer to a heated serving dish. Pour over the bolognese sauce. Serve immediately sprinkled with Parmesan cheese.

AFTER-HOLIDAY WINES

SPUMANTE

You've been away, had a great time and now you're home. Have a glass of wine, banish those post-holiday blues

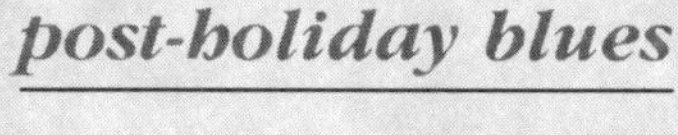

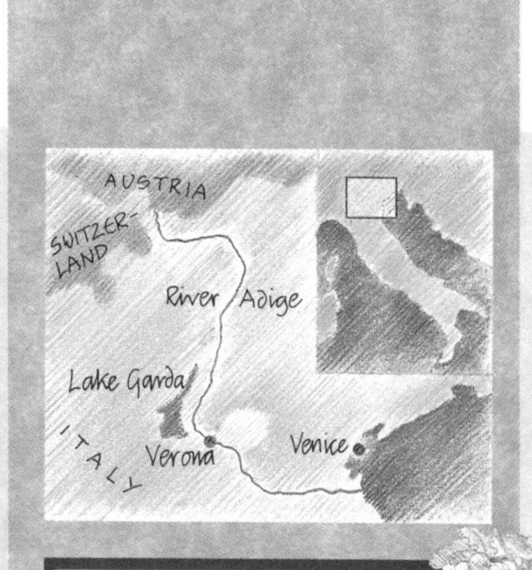

YOU KNOW THE feeling. You are back home after summer holidays. All is fine for the first couple of days while you are still relaxed and can't get too concerned about work, home, anything. Then suddenly it hits you: it is all over for another year. The carefree days are gone; the normal everyday problems start niggling again; the money is all spent; the days are getting shorter; the cold weather is approaching; and there is nothing more to look forward to until Christmas. It is all too easy for depression to descend like a thick fog. What to do? Well, one bottle of wine is not going to chase the blues away for good but it is amazing how a few little treats can start to make you feel that life isn't such drudgery after all and a bottle of wine can easily be one of those treats.

Cheer-up juice

If you spent less than you had expected on your sojourn and can afford a splurge, there is no better cheer-up juice than a bottle of Champagne – the more expensive the better. Underspending on holiday, though, is a rarity; the reverse more likely, so there's no point in adding guilt to the misery by paying too much for a bottle. Nevertheless, bubbles certainly have a remarkably uplifting effect; sometimes just the pop of a sparkling wine cork can make things begin to seem brighter.

Fun and frivolity ought to be the order of the day, so what better than a bottle with a name like **Peachy.** It is not just sparkling wine, although 99% of it is. The rest is from peaches, giving natural peach aromas and flavours, and even though it only makes up 1%, the wine is pervaded by a delicious peachy smell and taste. With all that peach around, it is needless to say that the wine is astoundingly fruity and noticeably refreshing too. The bubbles are not too strong; they just fizz gently making the drink even more restorative and enlivening.

Low-alcohol lifters

Not only might you want a cheer-up, you might well be feeling it is time to give your liver a rest; holidays can be quite boozy, or at least involve more drinking than you are used to. The two aims are not necessarily contradictory, at least not if you go for **Pétillant de Listel.** Not only has it blues-dispelling fizz and a soft, rounded grapey flavour, it also con-

BEST BUYS

Chardonnay Vino Spumante Santi
Taste guide: **dry**

Peachy
Taste guide: **medium-dry**

Pétillant de Listel
Taste guide: **sweet**

Domaine de Lafitte Rosé 1988 Vin de Pays des Côtes de Gascogne
Taste guide: **light-bodied**

Viña Albali Tinto Reserva 1983 Valdepeñas
Taste guide: **medium-bodied**

tains only 3 per cent. alcohol. It comes from the south of France, near Montpellier, from the impeccable Listel company, who have succeeded against all odds in growing good healthy grapes in sand by the salt marshes along the coast. After picking the Muscat, Ugni Blanc and Clairette grapes they stop•the fermentation just before it reaches 3 per cent. So there is still some grapey unfermented juice, but plenty of real wine character too.

Sparkling wines, even without containing peach aromas or grape juice, often taste very different from the wine before it was made sparkling. This is usually deliberate and beneficial; deliberate, to create a nice, soft yeastiness; and beneficial, as many wines which make great sparklers are thin and sharp on their own. An exception is **Chardonnay Vino Spumante Santi.** It starts as a still wine from the superb Chardonnay grape variety and is then made sparkling with great care and delicacy to retain the wine's initial, wonderfully attractive character.

Chasing the blues

Wines to push aside depression do not have to be sparkling, of course, they can be still too. Good blues chasers are often pink. The colour is one positive aspect; the taste, with the crispness of a white and the fruit of a red, another. Particularly good for post-holiday miseries is **Domaine de Lafitte Rosé 1988 Vin de Pays des Côtes de Gascogne.** Its colour is quite a deep pink, almost a light red, so it doesn't look too flimsy for chilly evenings. It is dryish, with ripe fruit and a moreish bouquet and flavour of strawberries and roses. What could be more summery than that?

As the nights draw in and if the temperature is lower than you have become used to, a soft, comforting red may be your best bet. **Viña Albali Tinto Reserva 1983 Valdepeñas** is one such and coming from Spain, it is especially good for anyone who is just back from holidaying. It has an inviting bouquet that is smoky and resembles pomegranates.

To recapture your holiday mood, cook something warming, yet summery: Italian rice with chicken and mushrooms, below, is a perfect dish to lift your spirits and remind you of balmy days abroad. Use Arborio rice to give it an authentic touch.

PERFECT PARTNERS

Italian rice with chicken and mushrooms

- ***Preparation: soaking mushrooms then 20 minutes***
- ***Cooking: 1 hour 5 minutes***

2tbls dried mushrooms
1 chicken breast, boned
salt and pepper
425ml/¾pt chicken stock
225ml/8fl oz dry white wine
50g/2oz butter
1 onion, sliced
50g/2oz prosciutto or cooked ham, sliced and cut into thin strips
3tbls Marsala
175g/6oz Italian or medium-grain rice
100g/4oz frozen peas
pinch each of dried marjoram, thyme and basil
2tbls grated Parmesan cheese, to serve

- ***Serves 4***

1 Soak the mushrooms in tepid water for 45 minutes. Meanwhile, season the chicken generously. Place it in a small saucepan and cover with 150ml/¼pt each stock and wine. Bring to the boil, reduce the heat and simmer gently, covered, for 25 minutes or until tender, turning over once during cooking. Skin and dice the chicken breast.

2 In a flameproof casserole, melt the butter. Add the onion and sauté for 10 minutes or until lightly golden. Drain and dry the mushrooms and add to the pan, together with the strips of prosciutto or ham and the chicken. Sauté gently for 2-3 minutes.

3 Add the remaining wine and the Marsala. Bring to the boil and sprinkle in the rice and peas. Simmer gently until the wines have been absorbed.

4 Add the remaining stock and the herbs and season with salt and pepper to taste. Simmer gently for 20 minutes or until the rice is tender but not mushy, stirring occasionally to prevent sticking. Just before serving, toss in the grated Parmesan cheese.

PROVENCAL PARTNERS

VIN DE PAYS DE L'AUDE

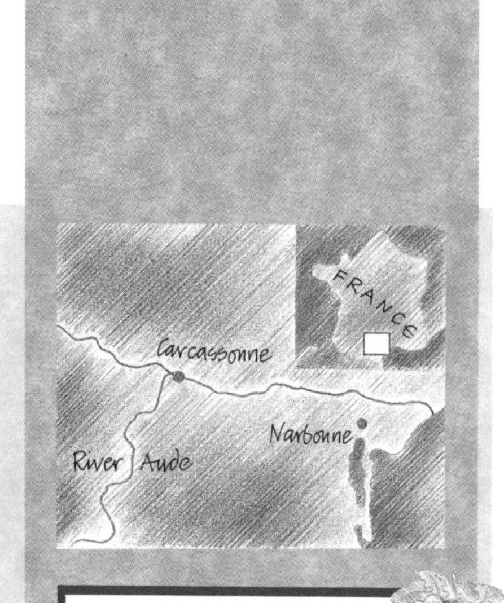

BEST BUYS

Vin de Pays de l'Aude Union des Grands Crus
Taste guide: light-bodied

Corbières 1988 Les Caves de Vignerons Embres
Taste guide: medium-bodied

Jeunes Vignes
Taste guide: very dry

Dom Hermano 1986 Reserva
Taste guide: full-bodied

Chianti 1987 Cantine Leonardo da Vinci
Taste guide: medium-bodied

Côtes de Provence
Taste guide: medium-bodied

The richness of ratatouille welcomes the red wines of the South

RATATOUILLE IS ONE of the most delicious mixtures of vegetables – the combination of the flavours of aubergines, onions, peppers, courgettes and tomatoes, all good, strong, flavours, and all held together and amalgamated by olive oil, can be remarkably moreish. Ratatouille may be eaten hot or cold. It is often served as a side dish, frequently with meat dishes, but is also great as a meal in itself: a great pile of colourful tastiness which needs only a hunk of bread and some wine to accompany it.

Local delights

You would have thought that there would be dozens of wines that would go well with ratatouille. After all, the ingredients are typical of Provence, which is an important wine-producing region: there should be a local match. Yet it is hard to partner ratatouille really well, for the natural juices of vegetables like peppers and aubergines can react badly with wine and onions are particularly difficult. The result is that the wine can seem less fruity than it is, perhaps a bit vinegary and maybe hollow too. However, the simple, light, quaffing wines which dominate the south of France are some of the best bets for drinking with ratatouille. They keep their fruit and character and form the perfect base to let the strong flavours of the food remain the dominant factor. One of the best is **Vin de Pays de l'Aude Union des Grands Crus**, which comes from the Mediterranean south of France, near Narbonne and Carcassonne. Remarkable value and a bright, purplish ruby colour, it is uncomplicated to drink, with its easy, light, dusty, lively but fresh fruit and its soft, rounded style. It tastes even better with ratatouille and forms just the right sort of contrast and flavour to set the food off.

From much the same area also comes **Corbières 1988.** Try the one produced by Les Caves de Vigneron Embres. It is deeper in colour than the Vin de Pays de l'Aude, with a vivid blue-purple colour indicating its youth. It too has good, lively fruit, but with a little more weight and a brambly style. Unlike some stronger wines, which compete with the ratatouille rather than complement it, the Corbières is just the right weight to avoid affecting the vegetables' flavours.

Try white

Why not a white wine? It can be just the thing when ratatouille is served with white meats and sometimes when it is eaten cold. A really special (but not specially expensive) wine is **Jeunes Vignes**. The name means young vines and that is exactly what the wine is made from – grown in the Chablis area in the Yonne département of northern France, in vineyards known for the high quality of the wine. Once the vines are older, they will be used for Chablis production, but in the meantime Jeunes Vignes gives you the chance to drink wine almost as good as Chablis, but for a fraction of the price. Even more important, it goes beautifully with ratatouille, its creamy, slightly spicy but very crisp taste balancing the richness of the oily vegetable stew very well indeed and its peppery finish becoming even more apparent – almost as an extra seasoning to the dish.

Pepperiness is also the clue to the good partnership of **Dom Hermano 1986 Reserva** with ratatouille. It is a Portuguese wine, big, ripe and punchy with flavours of spice and leather and prunes as well as pepper. It would seem to be too fully flavoured for the food, but the flavour is in the richness of the fruit more than in sheer heftiness and, in fact, the combination works well. The wines seem a mite less fruit-packed with the food, but the ratatouille loses nothing.

Mediterranean mixtures

From another part of the Mediterranean, there is an intriguing mix between the lively, sharply fresh, spicy fruit of a young **Chianti 1987 Cantine Leonardo da Vinci** and the oily richness of the ratatouille vegetables. The Chianti cleans the palate excellently, leaving it fresh to enjoy the next mouthful, while itself being toned down by the food's flavours and textures.

For the most reliable match of all, go for a wine from Provence, ratatouille's homeland. **Côtes de Provence** may be white and is often rosé, but it is the red version that is best with ratatouille. Deeply coloured, with a powerful bouquet that recalls the warmth and herb-filled earth of the region, and a dry, rounded, ripe fruit taste, it is the right weight to balance this delicious mixture of vegetables.

Ratatouille, like many tasty stews, benefits from being left to stand so that the flavours can really develop. Leave to cool, then chill overnight. However, like a red wine, ratatouille shouldn't be served too cold. Let it warm to room temperature, or reheat gently over a moderate heat.

PERFECT PARTNERS

Ratatouille

- ***Preparation: 30 minutes plus 30 minutes standing time***
- ***Cooking: 40 minutes***

2 aubergines, quartered and cut into 5mm/¼in slices
salt
8tbls olive oil
1 Spanish onion, sliced
2 green peppers, seeded and cut into 2.5cm/1in squares
2 small courgettes, cut into 1cm/½in slices

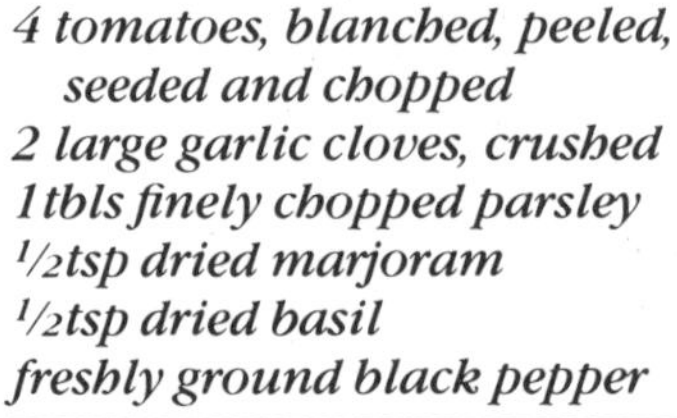

4 tomatoes, blanched, peeled, seeded and chopped
2 large garlic cloves, crushed
1tbls finely chopped parsley
½tsp dried marjoram
½tsp dried basil
freshly ground black pepper

- ***Serves 4***

1 Sprinkle the aubergine slices generously with salt. Place in a colander and leave for 30 minutes so that the bitter juices drain away. Rinse the sliced aubergine in cold water and then drain thoroughly on absorbent paper and pat dry.

2 Heat the olive oil in a flameproof casserole, add the sliced onion and sauté for 2–3 minutes or until transparent.

3 Add the green peppers and the aubergine slices, cover and cook the vegetables for 5 minutes.

4 Add the sliced courgettes, chopped tomatoes, garlic, parsley, marjoram and basil and season with salt and black pepper to taste. Cook, covered, for a further 30 minutes. Serve hot from the casserole, or cold as a starter for a summer meal. It is worth making a large batch as this dish freezes well.

AL FRESCO WINES

CASA PORTUGUESA

A picnic calls for very different wines, depending on whether it is an impromptu outing or a carefully planned grand occasion

IF THERE'S ONE thing nicer than a picnic, it is the anticipation of a picnic: imagining that perfect site, secluded, calm and quiet, sunny but with shade, insect-free; unearthing the hamper or cool-bags: preparing the food and utensils; wrapping up the wine. Of course, there are picnics and picnics. The sorts of food, and wines, you would pack for Glyndebourne wouldn't be the same as for a quick dash to the nearest grassy spot when the weather unexpectedly turns fine.

For the simplest impromptu meals out of doors, when the fun of it is the important thing and food is not the major consideration, there is

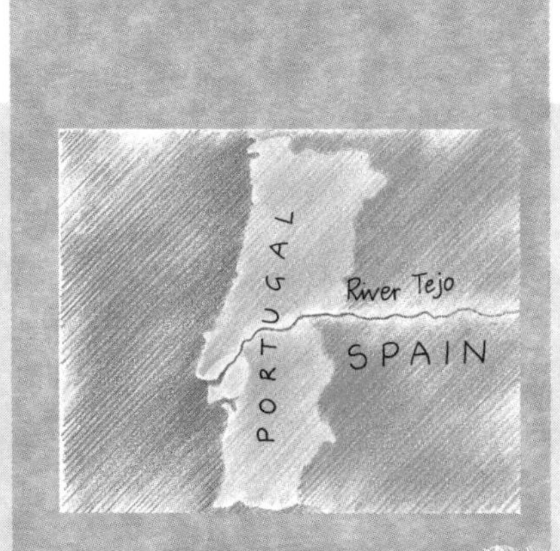

BEST BUYS

Casa Portuguesa
***Taste guide:* medium-bodied**

Muscadet de Sèvre et Maine 1987
***Taste guide:* very dry**

Pouilly Fumé Les Griottes 1987
***Taste guide:* very dry**

Cante Cigale 1987
***Taste guide:* dry**

Fleurie 1988 Georges Duboeuf
***Taste guide:* light-bodied**

Moscato d'Asti 1988 Gallo d'Oro
***Taste guide:* sweet**

KEEPING WINES COOL

The imagined picnic always has delightfully cool white (or rosé) wines. The reality is often quite different. For no matter what temperature a wine is when you leave home, it is all too easy to find the bottle nicely boiled when you come to drink it. The ideal picnic, of course, has a gently flowing stream close by. Should this be your fortune and you remember to keep some string in the car all you have to do is tie a long piece of string round the bottle (in such a way that it can't work loose) with a long end and dangle the bottle in the stream. The cold flowing water will cool it down quickly.

Cool bags work quite well as long as they are tightly packed, there are plenty of ice packs, and everything is well chilled when it first goes into the bag.

It is quite possible to keep bottles cool without buying a special bag. They must be very well chilled first (all night in the coldest part of the fridge). Then they should be wrapped in wet-damp towels. Any warmth then evaporates the dampness from the towels and doesn't reach the bottle. Wrapping the bottle in damp newspaper works just as well but can be messy (though it has the advantage of being disposable).

no point in worrying too much about the wine. Once, though, you find yourself planning the food with relish, it is time to think a little about the wine too. However good food tastes in the open air, the right bottles will make it taste even better.

White to picnic

White wine is many people's first choice for a picnic. It needs to be light, but flavoursome and crisp, so the Loire valley is a good place to start looking. A good Muscadet is ideal, like **Muscadet de Sèvre et Maine Sur Lie 1987 Domaine du Vieux Chais.** The Muscadet area is famous for its seafood and Muscadet is a perfect accompaniment. But even if your picnic isn't full of prawns, mussels and such like, this Muscadet, with its lively freshness and zingy acidity, will bring out the flavours of whatever you are eating.

For a really up-market picnic, a white wine to go for, also from the Loire, but this time from further up river, is elegant, elderflower-like **Pouilly Fumé Les Griottes 1987.**

If the idea of rosé seems right, there are few wines more suitable than **Cante Cigale 1987** from the Grenache grape with its salmon colour and fresh-fruit flavours.

Red range

A red wine is just as good a choice for a picnic. It is less sensitive to temperature, so if the cooling arrangements don't work too well it will still taste good. It is also the best accompaniment to cold meats and salami, some cheeses, and many of the more substantial contents of hampers. Whether you decide to take just red, or some red and some white, the red shouldn't be too full-bodied or tannic: fruity easy-drinking reds will taste fine on the picnic site; heavy, structured wines will taste odd. Try **Casa Portuguesa.** It is from Portugal, is firm enough to cut through any rich or fatty foods, but is smooth tasting and has enough succulent fruit to pep up even the most ordinary foodstuffs and cheer the drinkers.

The posh picnic could instead be graced by a posh Beaujolais: try **Fleurie 1988 Georges Duboeuf.**

Extra fizz

What about fizz? Well, for Glyndebourne nothing less than Champagne will do. Generally, though, being bumped around in the back of a car makes sparkling wine pretty explosive and there is no point in buying an expensive bottle, only to waste half all over the ground. A good compromise is a lightly sparkling wine, light enough to come with an ordinary (not mushroom) cork. It will certainly be fizzier than normal at the end of its journey, which is just what you want anyway. **Moscato d'Asti 1988 Gallo d'Oro** is just right – light in weight, light in alcohol and deliciously grapey.

The food for a picnic, no matter how posh, needs to be transportable too, and the picnic pies, below, are perfect. You could vary the filling to suit the wine, adding canned fish or chopped cooked chicken in place of the meat, if you prefer.

PERFECT PARTNERS

Picnic pies

- ***Preparation: making pastry, then 40 minutes***
- ***Cooking: 20 minutes***

700g/1½lb ready-made shortcrust pastry, thawed if frozen
8 anchovy fillets, finely chopped
1 small onion, finely chopped
1 garlic clove, finely chopped
4tbls olive oil
2tbls finely chopped parsley
225g/8oz cooked ham or veal, finely chopped
1 egg, separated, plus 1 yolk
salt and pepper

- ***Makes 16***

1 Heat the oven to 190C/375F/gas 5. Roll out the pastry 3cm-6cm/⅛in-¼in thick and cut out 32 circles using a 7.5cm/3in biscuit cutter. Use a small decorative cutter to remove a small shape from the centre of half of the circles.

2 Put the anchovies, onion, garlic and oil in a mortar and pound to a smooth paste. Transfer to a bowl.

3 Using a wooden spoon, beat in the parsley and ham or veal, adding just enough egg yolk to bind. Season to taste.

4 Place about 1tbls of the mixture in the centre of each of the whole pastry rounds. Brush the edges of the dough with egg white and cover with the remaining rounds, pressing the edges well together and fluting them with the back of a knife. Brush the tops with beaten egg yolk and bake for 20 minutes or until golden. Serve warm or leave to cool.

COOL REDS FOR HOT WEATHER

TEROLDEGO

Serve chilled red wines on warm summer days – you'll find them refreshing, reviving and deliciously different

THERE IS ALWAYS at least one spell of good summery weather each year, even thougth it can be easy to forget about it after a batch of typical British gloom. Once the sky turns blue and the sun starts shining it is time to stock the fridge with plenty of cooling white wine. Or is it? Why not red? The reputation of red wine may be as something fuller, heavier and more suitable for coolish weather and the heavier food that goes with it. Yet there are plenty of red wines that are light enough for the hottest of days, that taste just as good, if not better, with a light chilling, and that have more mouthwatering fruit than their white counterparts.

Sunshine reds

One of the best examples of the sort of red wine that is great when the sun is shining is Beaujolais. Beaujolais' great advantage is the grape from which it is made: the Gamay. This gives it lightness, great drinkability and, best of all, a wonderful fruitiness that zings right through the wine – both on the bouquet and the taste. There are some Beaujolais that are a little, or a lot, weightier and richer and which go by names like Fleurie, St Amour or Moulin à Vent. But for hot-weather drinking there is no need to splash out on these. However, for just a little extra fruit than straightforward Beaujolais, look for Beaujolais Villages. This can come from one or several of a number of villages in the Beaujolais region of France, which produce particularly good wine. A delightful example is **Beaujolais Villages 1987** bottled for Paul Boutinot. It is at its best a little on the cool side and served with cold chicken dishes, especially those with fruity sauces or relishes.

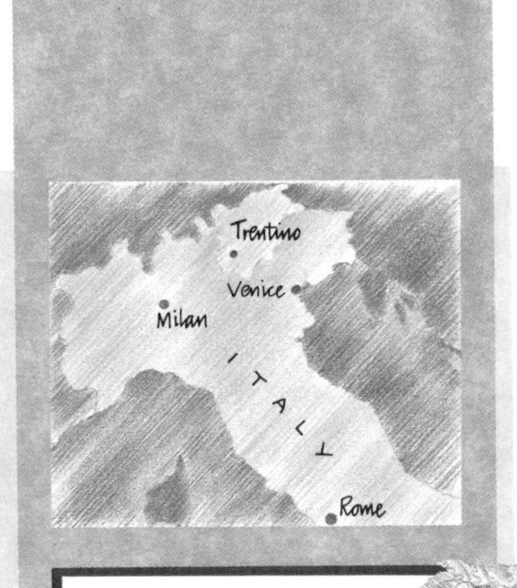

BEST BUYS

Teroldego Rotaliano 1987 Gaierhof
***Taste guide:* light-bodied**

Beaujolais Villages 1987
***Taste guide:* light-bodied**

Domaine des Acacias 1987 Touraine Cabernet
***Taste guide:* light-bodied**

Château la Gordonne 1986 Côtes de Provence
***Taste guide:* light-bodied**

Rheinpfalz 1982er Rotwein
***Taste guide:* light-bodied**

Light and lively

The Gamay grape thrives further north in France too, in the Loire valley. But as well as Gamay, there is also Cabernet. Usually, this is not the blackcurranty Cabernet Sauvignon which is widespread all over the vine-growing world, but Cabernet Franc, which produces a lighter, livelier, grassier wine. This liveliness and grassiness is especially pronounced in the Loire's Touraine where **Domaine des Acacias 1987 Touraine Cabernet** comes from. Its herbaceous bouquet and quite noticeable wood fruits flavour would suit many occasions – even when the weather isn't quite so good. When it is slightly chilled, though, the weight of the wine seems much less pronounced; its fruit and grassiness is accentuated. Try it, particularly, with cold ham or with pork dishes.

Full of good cheer

Another wine that should convert you to the idea of drinking red in summer is **Teroldego Rotaliano 1987 Gaierhof.** It comes from the Rotaliano plain of north east Italy's Trentino province, which is unusual in itself as most wine production of decent quality comes from slopes, not flat land. Even more unusual, though, is the grape used, Teroldego. This not only thrives practically nowhere else, but doesn't even produce particularly well on the hillsides near the Rotaliano plain. It can make big wines that take several years to become ready for drinking. It also, however, often produces wines that are at their best two to three years after the harvest. The wine is deeply coloured, and a dark reddish purple, which makes you think it is going to be much weightier and fuller-bodied than it actually is. Its crisp, herby, spicy, blueberry bouquet leads on to a delicious, soft, fruit-packed mouthful, with plenty of life but not much tannin, which is equally good at room temperature or a little cool. It is versatile enough to accompany most meat dishes, quiches and other cheesy dishes, vegetables and salads.

Although Provence in the south of France is best known for its rosés, a lot of red wine is produced (and consumed!) there, so there should be little doubt about its suitability for hot-weather drinking. **Château La Gordonne 1986 Côtes de Provence** is a particularly soft, attractive, well-balanced example. As with all hot-weather reds the fruit is the dominant factor, but here the fruit character is quite soft and ripe and the wine is gentle and easy drinking, but doesn't lack flavour. Just like most of the reds mentioned it could be served lightly chilled, but it is better at room temperature so is best chilled in the fridge first only if it is to be drunk out of doors where it would warm up quickly. It is a natural partner for provençal foods, and light meals such as, for example, baked potatoes served with hot or cold ratatouille.

Rich and fruity

Germany produces some red wines too, even though most of its production is white. Often the reds are pale in colour and very light in style. Not so **Rheinpfalz 1982er Kapellener Kloster Liebfrauenberg Dornfelder Rotwein Trocken.** The name may seem cruelly long, but is information-packed. For example, Rheinpfalz is the region the wine comes from, Dornfelder the grape variety, Rotwein indicates it is a red wine, and Trocken shows that it is dry. It has a vivid purple-red hue, smells of blackberries and raspberries and has the rich flavour of summer pudding. As with the other light reds in this section, this wine is at its best served slightly chilled instead of at room temperature. It is good with summery foods, such as lamb and pepper kebabs, below, which, if the weather happens to be summery too, can be barbecued outside rather than grilled indoors.

PERFECT PARTNERS

Lamb and pepper kebabs

Serve these colourful kebabs with pilaff and a crisp green salad

- ***Preparation: 20 minutes, plus marinating***
- ***Cooking: 20 minutes***

700g/1½lb boned leg of lamb
3 sweet peppers, red, green and yellow, seeded and cut into chunks
8 bay leaves
For the marinade:
3tbls oil
3tbls wine vinegar
1 garlic clove, crushed
¼tsp mustard powder
1tbls chopped fresh mixed herbs (parsley, rosemary and thyme)

- ***Serves 4***

1 Blend together the marinade ingredients. Put the meat in a large shallow dish and pour the marinade over. Leave for at least 2 hours in a cool place.

2 Heat the grill to medium-high. Drain the lamb, reserving the marinade. Divide the lamb, peppers and bay leaves between eight wooden skewers.

3 Arrange the kebabs on the grid, brush with the marinade and cook for 15-18 minutes, turning over frequently and basting with the marinade. Serve hot.

BEST FOR BARBECUES

CANNONAU

Planning the weather may be out of your control – but make sure the wine you serve matches the flavours of charcoal-grilled food

BARBECUES ARE A great way to entertain, and an easy way to turn a simple family meal into a party – usually with a lot of keen helping hands to share the burden. Unfortunately, our unpredictable climate means that a barbecue in the back garden is probably a less common piece of equipment than a microwave in the kitchen. After all, there seems little point in spending ages coaxing the coals to heat up if the result is just a cold, uncomfortable meal, with the wind blowing smoke in your eyes and the food's flavour away, or worse – if it starts raining. For a family meal the temptation to take your plate and retire indoors can be very strong. Whereas if it's a party, battling against the weather becomes part of the fun, and if the weather's fine, then that's an added bonus.

Hence wines for barbecues are wines that will suit large gatherings just as much as the charcoal-grilled flavours of the foods. It's advisable to keep the wine indoors and serve people at regular intervals. This way it will stay at the required temperature and not boil if placed too close to the fire.

Sardinian star

Unless it's a seafood barbecue you are planning, reds are best. An ideal red is one that is fresh and fruity enough to attract white-wine drinkers; full and flavoursome enough to match the smells and tastes of broiling chunks of meat; light enough to be enjoyable throughout the evening, without palling; and not too expensive as you may be catering for a large number of people. It seems a tough order to fulfil, yet there's a light-bodied red from Sardinia that does: **Cannonau del Parteolla 1986**. Sardinia is traditionally known for strong, beefy wines – but then producers discovered modern ways of making wines and realized how good their grapes were for such styles. Cannonau, the grape variety, is widely planted in Sardinia and, despite the very different-sounding name, is the same as the Grenache grape of the Rhône valley.

Tasting Cannonau del Parteolla

BEST BUYS

Cannonau del Parteolla 1986 (1 glass)
Taste guide: light-bodied
Serve with: all barbecued food

Côtes de Provence 1986 (1 glass)
Taste guide: medium-bodied
Serve with: steaks, chops and kebabs

Côtes du Rhône (IL) (2 glasses)
Taste guide: medium-bodied
Serve with: steaks, ribs

Orlando Jacob's Creek Dry Red 1986 (1 glass)
Taste guide: full-bodied
Serve with: red meats

Orlando Jacob's Creek Dry White (1 glass)
Taste guide: dry
Serve with: chicken, fish

has the same effect as biting into a ripe English plum – the flavour is pronounced and it has a mouth-watering, sweet-sharp character. It is classed as a 'low tannin' wine which means the mouth isn't dried out when you drink it (so white wine drinkers won't dislike it) and the flavour lingers.

A touch of Provence

If it's summertime, and especially if the weather is behaving itself for once, there can be few wines more apt than Côtes de Provence. For anyone who has holidayed in this part of southern France, merely the sight of its tall curvaceous bottle should evoke hot days, balmy nights and *al fresco* meals.

Much Côtes de Provence is rosé. There's also plenty of red, though, like the medium-bodied **Côtes de Provence 1986** bottled by Les Maîtres Vignerons de St Tropez. A good deep purple colour, it makes a firm impression in the mouth with a silky, ripe, crisp fruitiness and, some say, the taste of Provençal herbs. Stretching your imagination even further, you can almost sense charcoal smoke too.

A barbecue party is a good occasion to serve wines in larger bottles – not the huge 2-litre bottles which are heavy and difficult to pour, but 1-litre bottles (or at a stretch 1½ litres). A good buy is **Côtes du Rhône** bottled by Delaunay l'Etang-Vergy. There are many examples of Côtes du Rhône available, and often they are quite distinct from each other in style. For Côtes du Rhône can be made from any one or any combination of a wide variety of grapes (Grenache is the most common) and by a variety of wine-making techniques. This example has all the characteristic pepperiness of good Rhône wines but isn't as intensely flavoured as some, nor as light as others – so it's perfect drinking with assertively flavoured barbecued food but can be quaffed on its own quite happily too.

'Aussie's' choice

So natural is barbecueing to the Australian lifestyle, it would be surprising if Australian wine didn't go well with it. Most does. But for something that suits most people, even if temperatures are below the 80s, try **Orlando Jacob's Creek South Australia Dry Red 1986**. Dry, but with a rich fruitiness like crushed loganberries, as it hits the mouth it just bursts into flavour – and will make everyone ravenous.

If you're barbecuing chicken, fish or vegetables, or if someone prefers not to drink red, then there are plenty of charcoal-matching whites from which to choose. For example, there's the well-rounded and fruity **Canterbury California Sauvignon Blanc 1986** (see page 48), **João Pires 1987**, an excellent Portuguese wine that is perfect for drinking outdoors (see page 124), and **Morio-Muskat 1986** (see page 56) which comes from Germany and has plenty of character – it is lightly aromatic, fruity, and has a touch of spice to boot. Or, why not go for the Australian red's white partner, **Orlando Jacob's Creek South Australian Sémillon Dry White**? Its dry, lemony fleshiness is just what is needed to complement the tasty vegetable kebabs, below.

PERFECT PARTNERS

Vegetable kebabs

- ***Preparation: 15 minutes, plus marinating***
- ***Cooking: 25 minutes***

6 baby turnips
2 courgettes, cut into 12 × 2.5 cm/1 in pieces
12 large button mushrooms
6 tinned baby sweetcorn, cut in half
6 small tomatoes, cut in half
6 stoned olives
For the marinade:
1 small onion, very finely chopped
1 clove garlic, crushed
50ml/2fl oz red wine vinegar
50ml/2fl oz olive oil
1 tsp soft dark brown sugar
1 tbls Worcestershire sauce
3 tbls tomato ketchup
1-2 drops hot pepper sauce
1 bay leaf
salt and pepper

- ***Serves 4-6*** 

1 Place the peeled turnips in a saucepan of boiling water. Boil for 8-10 minutes or until tender. Rinse, then drain.

2 Plunge the courgettes into a saucepan of boiling water for 1-2 minutes to blanch. Rinse under cold water, then drain.

3 Mix all the marinade ingredients together in a large shallow dish. Add all the vegetables, toss well, cover and leave to marinate in the fridge for 1-2 hours, stirring two or three times.

4 Thread the vegetables onto six metal barbecue skewers. Place over hot coals for 5-7 minutes on each side, turning and basting with any remaining marinade.

WINES TO DRINK IN THE GARDEN

PINOT GRIGIO

Wine drunk indoors may not be the best choice for casual sipping in the garden. Choose a wine with character, then sit back and relax

1987
PRODUCE OF ITALY
PINOT GRIGIO
VINO DA TAVOLA
DEL TRIVENETO
SELECTED BY SAFEWAY BOTTLED BY G.I.V. S.p.a. PASTRENGO

ON THOSE ALL too rare days when the weather is balmy and you have time to enjoy it, it's wonderful to sit in the garden and relax with a glass of wine. To get the most out of your glass it is best to choose a wine that is fairly resistant to temperature changes, that is, one that doesn't lose all its flavour if chilled too much, or that becomes dull and flat if it gets too warm. For it is surprising how quickly wine straight from the fridge will warm up in the sun, and even more surprising how cold it can get as the temperature becomes cool at dusk.

Blowing in the wind

Remember too that there's often a breeze outside which can blow away all the aroma of a wine with a delicate bouquet. It's better to choose either a medium- to full-bodied wine with lots of aroma, which will keep its perfume whatever the wind does, or a lightish wine without much bouquet to lose.

Of the latter, despite the fact that Italian wines are made to be drunk with food, many of the whites, like a good Soave, Frascati or Orvieto Secco, are excellent garden wines. For the ideal quaffer on a hot day there's **Vermentino di Sardegna 1987** produced by C S di Dolianova. Sardinia (the Italians spell it Sardegna) used to be known for big, heavy, alcoholic wines. But this white Vermentino is the complete opposite. Produced using the most advanced techniques for obtaining freshness and delicacy from grape juice, the wine is as crisp and refreshing as you could wish. And with lowish alcohol (10.5%) it is just right for outdoors: sitting in the sun it is easy to drink more than you realize and not feel any effect until you go back indoors – by which time it is too late!

For a relaxing glass with family or friends, either before (or with) a picnic lunch or at the end of the day, another Italian wine is perfect. It's **Pinot Grigio 1987 Vino da Tavola del Triveneto** and it fulfils all the criteria for a fresh-air wine. You could drink it ice cold or almost up to room temperature, in a breeze or on a windless day, and its characteristic taste would still give pleasure. It comes from the north east of Italy and is made from a grape variety, Pinot Grigio, which is perfectly suited to the climate and soils there and produces flavoursome wines of character. Gently silky on the nose, its taste is rich and penetrating,

BEST BUYS

Pinot Grigio 1987 Vino da Tavola del Triveneto
Taste guide: dry

Vermentino di Sardegna 1987
Taste guide: very dry

João Pires 1987
Taste guide: dry

Albiger Petersberg Auslese 1985
Taste guide: medium-dry

Jekel Vineyard White (Johannisberg) Riesling 1986
Taste guide: dry

rounded and not too acidic, but firm and clean-feeling.

Grape greats

Muscat is one of the most adaptable of grape varieties, producing anything from the lightest of wines to the fullest, and from ultra-dry to very sweet, yet it always retains its muscatty aroma – just like biting into fresh Muscat grapes. Many countries grow the grape and there are stylistic differences, 'national character', from one to the next. Producer of garden Muscat *par excellence* is Portugal with one wine, **João Pires 1987**. The firm, grapey bouquet is so penetrating it would take a really strong wind to blow it away; the taste is ripe, also grapey, but with all sorts of other fruits noticeable too, both green and tropical, which makes it taste like a liquid fruit salad.

Sweetness and light

On that wonderful balmy evening, as the sun sets through a cloudless sky turning from blue to pink to purple, a gentle, refined glass of something elegant from Germany is hard to beat. For this sort of occasion it is worth getting something just a little fuller and a touch sweeter than usual, like **Albiger Petersberg Auslese 1985** from Ewald Theod Drathen. The important word here is Auslese. It means that selected bunches of particularly ripe grapes have been used – but instead of all that extra grape sugar being turned into alcohol, the alcohol level remains quite low (9.5%) and a little sweetness remains. The delicate, really floral aromas on this wine are well worth savouring, so it's best when the air is still.

For just as much aroma, but more punch, try **Jekel Vineyard White (Johannisberg) Riesling 1986**. Despite the words Johannisberg and Riesling in the name, this wine is Californian! Its name emphasizes that the wine is made from the real Riesling (or white Riesling) grape, which comes from Germany, not any other related variety. The nose is like honeysuckle and sweet lemon, while the taste is so floral and fruity the wine seems almost sweet. But it isn't, it's dry, and with a mouthfilling zingy fleshiness that demands to be noticed.

Rose tints

And don't forget pink wines like **Listel Gris de Gris** (see page 35). With their delightful eye-catching colour, not too much body, but with good structure, they are perfect for summer drinking and could certainly turn a relaxing afternoon in the garden into something worth remembering.

PERFECT PARTNERS

Italian tuna and bean salad

Eat this salad in your garden with a cool glass of Pinot Grigio

• ***Preparation: 10 minutes, plus 1 hour marinating***

400g/14oz canned cannellini beans
2tbls finely chopped onion
2tbls finely chopped parsley
1 garlic clove, finely chopped
100g/4oz canned tuna fish, drained and broken into chunks
onion rings, black olives and watercress sprigs, to garnish
For the dressing:
6tbls olive oil
2tbls wine vinegar
½tsp Dijon mustard
salt and pepper

• ***Serves 4*** (£)

1 Drain the beans and rinse with cold water. Drain again and toss gently in a bowl with the finely chopped onion, parsley and garlic.

2 Beat the oil, vinegar and mustard together with a fork until they form an emulsion. Add salt and pepper to taste. Pour this over the beans and toss gently. Allow the beans to marinate in this dressing for at least 1 hour, turning carefully once or twice.

3 Using a slotted spoon, transfer the beans to a serving dish. Arrange tuna chunks down the centre of the dish and garnish with onion rings, black olives and watercress sprigs.

HEATWAVE WHITES

SYLVANER

Savour the magic of hot summer days with a delicious glass of chilled, crisp white wine

DESPITE THE DREARINESS of recent summers, we usually get at least one spell of sizzling weather sometime between spring and autumn. And who knows . . . apart from the fact that we must be due for a good summer at long last, one result of the greenhouse effect is that we just might be seeing a lot more heatwaves. Once the sun does start beaming there are few things more enticing than a glass of delicious white wine, chilled so well that beads of condensation start forming on the glass.

Quenching the thirst

If you're really hot, though, when you succumb to the delights of Bacchus the water content of the wine will be quickly absorbed to replace your body's natural loss. What remains to be propelled round your blood stream, therefore, will be comparatively stronger in alcohol, which is why drinking in the sun can make you tipsier than you expect. One way round the problem is to ensure you drink at least one tumbler of water for every glass of wine – and always have some water first if you're thirsty. Another solution is to choose a wine with a lower than average alcohol content – and German wine is the obvious choice. It needn't be too posh a bottle but it must have enough body to keep some flavour even when icy cold. Kabinett wines are the ideal solution, being just a little way up on the quality scale, but far enough up to make a difference. One of the best value Kabinett wines is **Bereich Bernkastel Kabinett 1985** by Moselland. It is soft, floral and delicately fruity. The touch of sweetness is grapey, not sugary, and well integrated into the wine, giving delightful balance. And it's only 8% alcohol.

Full of fruit

Muscadet used to be the hot weather wine par excellence. Light, fresh and crisp, if not all that characterful, it was an ideal reviver. Now, though, its place has been usurped by Vin de Pays des Côtes de Gascogne. Gascogne is just as light and refreshing but, being made from the Ugni Blanc and Colombard grapes and coming from further south in France, it has far more fruit. With its good herbaceous, appley quality and mouthwatering taste, it's a terrific cooler and enlivener. It is widely available. One of the best is called **Colombelle.** It comes just from the Colombard grape and is even more

BEST BUYS

Sylvaner 1987
Taste guide: **dry**
Serve with: **most light, cool, summery foods**

Bereich Bernkastel Kabinett 1985
Taste guide: **medium dry**
Serve **on its own**

Colombelle
Taste guide: **very dry**
Serve with: **light stir-fries**

Soave Classico 1988
Taste guide: **dry**
Serve with: **seafood salads**

Frascati Superiore Secco 1987
Taste guide: **dry**
Serve with: **vitello tonnato, anything Italian**

lively in flavour than the rest of the Gascogne wines.

Another way to eliminate the excessive effect of the alcohol is to put some ice cubes in your wine. This also has the benefit of preventing it warming up in your glass too quickly. Wine connoisseurs may frown at this sort of thing but it really does no harm at all – as long as the wine has enough substance to cope with its dilution and severe chilling. It is surprising how some wines, which don't appear to be excessively strongly flavoured, react admirably well to this sort of treatment. One such is **Soave Classico 1988 Zenato.** A young wine, it has an excitingly racy, zingy character overlying the more creamy-nutty flavours, but there's still the typical slightly bitter, almondy taste that you notice just after you swallow. Soave Classico has more concentration (without too much body) – enough to keep giving you flavour even though you may 'mistreat' it.

Another Italian wine, **Frascati Superiore Secco 1987 G.I.V.,** has exactly the same benefits. Perhaps being Italian is an important reason; the wines have to be suitable for drinking in hot weather, otherwise what would Italians do during their terrific summers? Frascati comes from the hills around Rome and is the wine the Romans drink on practically every occasion, with nearly all their most typical dishes. They drink it just lightly chilled, or icy cold; they may drink it full strength, they may add water. Whichever way, it always tastes just right. From the Trebbiano and Malvasia grape varieties, it is more strongly almondy than Soave, firmer, steely clean, and with herby fruit whose flavour persists in the mouth.

Superb Sylvaner

Probably the ideal for hot weather drinking, though, is the fascinating mixture of a German grape variety, with all its perfume and grapey aroma, but grown in France, where the grapes ripen more fully, wines are more assertive and are made totally dry. **Sylvaner 1987** Vin d'Alsace Les Viticulteurs des Caves Ernest Wein is the perfect heatwave bottle. Sylvaner is one of the most delicate and flowery of grape varieties – its bouquet is just like a sun-drenched field full of wild summer flowers. It tastes quite full, quite rich, but still fruitily floral. It's remarkably refreshing too, but impressive enough for more thoughtful sipping in the shade. There could be no better way to commemorate the first really hot day of summer – whenever it arrives.

And to eat with this sun-drenched wine, choose a summery dish such as monkfish salad (see below). Fresh herbs make all the difference to the flavour, so substitute fresh parsley rather than resorting to dried herbs.

PERFECT PARTNERS

Monkfish salad

- ***Preparation: 20 minutes, plus marinating and chilling***
- ***Cooking: 15 minutes***

1 fennel bulb
1 fish stock cube
350g/12oz tail piece of monkfish
1 large lettuce heart
stuffed olives, to garnish
For the dressing:
4tbls olive oil
2tbls lemon juice
1 garlic clove, crushed
½tsp chopped fresh tarragon
½tsp chopped fresh basil
salt and pepper

- ***Serves 4***

1 Trim the base and fronds from the fennel. Slice the bulb thinly and place in a bowl. Blend together the dressing ingredients and pour over the fennel. Leave to marinate for 2 hours.

2 Meanwhile, dissolve the fish stock cube in 600ml/1pt boiling water. Allow to cool slightly. Place the monkfish in a saucepan and pour the stock over the fish. Bring slowly to the boil, cover, reduce the heat and simmer for about 10 minutes or until the flesh just flakes. Remove the fish with a slotted spoon; leave to cool.

3 With a sharp knife, cut the cold fish into 4cm x 1cm/1½in x ½in pieces and toss with the fennel and the dressing. Chill, covered, for at least 30 minutes.

4 Finely shred the lettuce heart and arrange around the edge of a serving dish. Pile the fish and fennel in the centre and decorate with olives. Serve.

COTES DU LUBERON

Given the sheer range available, deciding on one or two store-cupboard reds sounds a tall order. Here's how to narrow the choice

IT'S A GOOD IDEA to keep a few bottles of wine at home; a full wine rack is probably as useful as a full spice rack. If someone calls round out of the blue, if you get home too late to buy wine for a dinner party, or if you suddenly decide you fancy a tipple with your meal, there's nothing easier than to reach out for a bottle and pull its cork, especially if it's a wine you already know you like.

The only disadvantage of having wine at hand is the possible incentive to drink it just because it's there! You'll probably find, though, that the more you get used to wine as a store-cupboard item, the less you are tempted to drink it for the sheer sake of it.

The ideal is to have one or two 'standby' wines that will suit a wide variety of occasions, that you will never fail to enjoy, and that are easy on the pocket. Once confident of their appeal, you can buy them by the case (12 bottles) and so take advantage of the case discounts many outlets offer. With reds the choice of possible standbys is wide. There are literally hundreds of good-value, good-tasting wines around; it's simply a case of discovering which style best appeals.

If you usually drink light-bodied wines when entertaining casually, on their own, or with light-weight dishes, a light-bodied standby is probably your best bet. For these, the best place to start looking is southern France, particularly the south west. Many such wines come into the category *Vin de Pays* but Fitou is a full *Appellation Contrôlée*, long recognized for the reliability of its quality. It comes from land behind the Mediterranean coast of France, south of Narbonne.

Fitou 1987, bottled by Sarl Bouffet, should soon become a firm favourite. A bright, ruby red, it has a soft nose of redcurrants and herbs. Soft in the mouth, without too assertive a flavour, its gentle, dry, smoky, herby taste will rarely seem out of place, either on its own or with a wide range of dishes.

Medium-bodied

Most people will probably plump for standbys somewhere in the medium-bodied category — neither too light nor too full. Italy has lots to offer in this category, as does southern France, but further to the east this time.

BEST BUYS

Domaine des Baumelles 1987 Côtes du Lubéron
Taste guide: medium-bodied

Fitou 1987
Taste guide: light-bodied

Gallaire et Fils Claret
Taste guide: medium-bodied

Barossa Valley Estates Shiraz Cabernet 1985
Taste guide: full-bodied

Oriahovitza Cabernet Sauvignon 1981
Taste guide: full-bodied

One of the most cheering wines is **Domaine des Baumelles 1987 Côtes du Lubéron**. Côtes du Lubéron is not far from Côtes du Rhône and has been declared a *Vin Delimité de Qualité Supérieure* – like a halfway house between Vin de Pays and Appellation Contrôlée. A good depth of colour, and a wonderful minty, spicy, blackberry nose, immediately set the expectations high. They are not disappointed. It tastes just like it smells, but more so, with plenty of freshness to give it liveliness. It's very easy to get hooked on this wine but difficult to get tired of it.

Bordeaux can also be a source of good, medium-bodied standbys, especially the soundly made wines often called simply Bordeaux or claret. These are usually blended from wines made in different parts of the region to ensure a characteristic, reliable flavour whatever the vagaries of the weather.

For a youthful, distinctively flavoured wine that is absolutely typical of classic Bordeaux flavour, try **Claret** produced by Gallaire et Fils. It's deeply coloured, it smells ripely blackcurranty and cedary, and it's firmly but attractively flavoured. Better with food than on its own, it's flexible enough to cope with an enormous variety of dishes, from shepherd's pie to tournedos.

Full-bodied

For the gutsier, fuller-bodied wines, it's worth looking further afield, particularly to Australia where so many outstanding wines have recently come from. Australia's more expensive wines are so concentrated and full bodied, they are probably too much of a good thing for everyday use. Thank goodness there are some in a lower price bracket which are just as toothsome but less intense.

Barossa Valley Estates Shiraz Cabernet 1985 is delicious. From Australia's cleverly invented blend of Shiraz and Cabernet Sauvignon grapes – both good at making exciting tastes – the wine is deeply coloured, and has a fascinating nose of plummy, curranty fruits and some tropical fruits, together with lemons and limes. Tasting ripe and rich, with sweet spiciness and enough acidity for freshness, this ought to be guaranteed to put you in a good mood and enliven your more wholesome dishes.

Perhaps the most obvious choice for a wine staple is Bulgaria and its Cabernet Sauvignon. For those who prefer more body, there's one of Bulgaria's top wines, **Oriahovitza Cabernet Sauvignon 1981**. From the Stara Zagora region, the wine is matured in oak casks for at least three years before bottling and is therefore permitted to be called Reserve. The oakiness is apparent on both nose and palate but it is combined with the gentle, velvety fruit that is the hallmark of Bulgarian wines, although with more concentration than usual. Try it with any meat to enjoy it at its best. It is perfect to serve with warming winter casseroles, roasts, and grilled or pan-fried steaks - you'll find that it really is too robust to serve with anything else. To drink with lighter main courses that you would serve for lunch or supper, such as the potato and Swiss cheese quiche, below, make sure that you have one of the lighter, medium-bodied red wines in store.

Potato and Swiss cheese quiche

- ***Preparation: 25 minutes***
- ***Cooking: 55 minutes***

225g/8oz shortcrust pastry, defrosted if frozen
flour, for dusting
butter, for greasing
300g/11oz potatoes, grated
50g/2oz Emmental cheese, grated
2 large eggs
150ml/¼pt single cream
salt and pepper
grated nutmeg
1tbls grated Sbrinz or Parmesan cheese

- ***Serves 4-6***

1 Heat the oven to 200C/400F/gas 6. Roll out the pastry on a lightly floured board. Grease a 20cm/8in flan case and line it with the pastry.

2 Line the pastry case with foil and beans and bake for 10 minutes. Reduce the oven heat to 180C/350F/gas 4, remove the foil and beans and bake for a further 5–10 minutes.

3 Combine the grated potatoes and Emmental cheese. Beat the eggs with the cream and stir into the potato and cheese mixture. Season well with salt, pepper and nutmeg to taste. Pour the mixture into the prepared pastry case.

4 Sprinkle the top of the quiche evenly with the grated Sbrinz or Parmesan cheese. Bake for 30–35 minutes or until the filling is set and the top golden brown. Serve hot.

VEGETABLE COMPANIONS

COTES DU RHONE

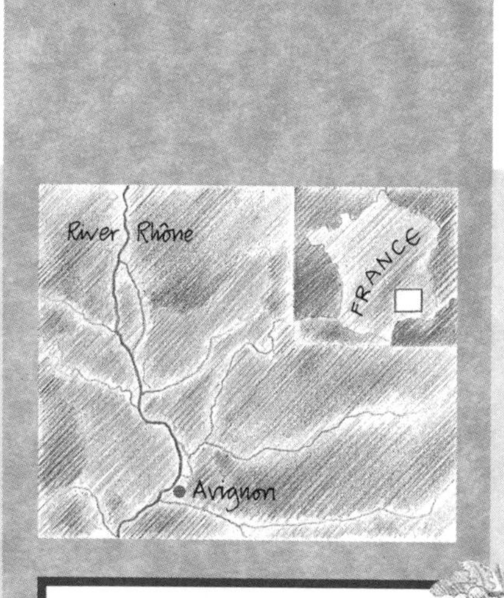

BEST BUYS

Domaine Saint-Apollinaire Côtes du Rhône 1987
***Taste guide:* full-bodied**

Domaine Anthea 1987 Vin de Pays d'Oc Cépage Merlot
***Taste guide:* full-bodied**

Fronton 1987 Domaine de Carmantran
***Taste guide:* full-bodied**

Some red wines from southern France make good companions for warming winter vegetable casseroles

WHEN WINES ARE rated for their ability to partner foods, it is usually the various types of fish, poultry, meat, game and cheese that are cited. Vegetable dishes are often forgotten. Yet the vegetable-based meals are becoming more and more popular – and not just among committed vegetarians. It may be in part due to a desire to eat more healthily, it may be because vegetables in season make for inexpensive meals, it may be simply because more and more people are realising just how good vegetable dishes can taste. Whatever it is, there is no need for wine to be ignored, as if vegetables didn't constitute a 'real' meal, especially when there are some great combinations to be had.

Take a vegetable casserole, for example, particularly a fairly substantial one based on winter standbys of potato, carrot, onion and cabbage. There are a large number of wines, mainly red, which accompany it remarkably well. Finding the perfect (rather than just a very good) match for many dishes is rare, but for a vegetable casserole there is one: **Domaine Saint-Apollinaire Côtes du Rhône 1987**. The wine is pretty moreish on its own, with a clean, fragrantly spicy, peppery, warming aroma and a full-bodied, sharply-fruited, wild blackberry taste. With the food, though, it just leaps into perspectve and tastes three times as good. Not only that, but it makes the flavours of the vegetables seem purer, stronger and better and can turn a 'humble' casserole into something quite special. The wine is organic, made without recourse to any chemical fertilisers or insecticides in the vineyards and with the absolute minimum of treatments in its production. This, perhaps, is why it has such an impressively clean taste – it 'sings'. The Côtes du Rhône form a large area towards the southern end of the Rhône river, centred on Avignon. Wines vary greatly in quality and style as they can come from one or more of several grape varieties and various wine-making methods are used. This is one of the classiest.

Also from the south of France, but further south, around Montpellier, is **Domaine Anthea 1987 Vin de Pays d'Oc Cépage Merlot.** It is made from the Merlot grape variety, which is found most often in Bordeaux, but flourishes wider afield too, with a rich, sometimes tobacco-ey, succulent, dried fruits taste. Domaine Anthea is deeply coloured, with good Merlot aromas, but also with an appealing earthiness the French would call 'Goût de Terroir' and a firm backbone of enlivening acidity and tannin. The food softens the attack of the wine. Although it may be too weighty if your recipe is on the delicate side, it will be a good balance for most vegetable casseroles and it creates a good background to enhance the vegetable flavours. Take care when adding curry flavours to vegetables, though, as this will change the wine picture (see page 99).

For a richer, slightly softer wine, that still shows those vegetables to good effect, try **Fronton 1987,** from the vineyards of Domaine de Carmantran, north of Toulouse in south west France. It is made from the Negrette grape, which is local to the area, with a little Cabernet Sauvignon (native to Bordeaux) and Syrah (native to the Rhône valley). This combination works wonders. The wine's smell is like very concentrated black cherries with cherry kernels and it tastes rounded, ripe, a little like dark fruit cake (also with cherries!). It is good value, it enlivens the food, and it is a joy to drink with vegetable dishes, such as cauliflower, mushroom and oat casserole, below. This delicious recipe is quick to make using a simple sour cream sauce to coat the vegetables. Use broccoli in place of the cauliflower, if you prefer.

PERFECT PARTNERS

Cauliflower, mushroom and oat casserole

- ***Preparation: 15 minutes***
- ***Cooking: 50 minutes***

1 medium-sized cauliflower, in even-sized florets
salt
freshly ground black pepper
2tbls flour
300ml/½pt soured cream
1tsp mild mustard
175g/6oz matured Cheddar cheese, grated
225g/8oz button mushrooms, trimmed
25g/1oz butter
100g/4oz rolled oats
50g/2oz coarsely chopped walnuts

- ***Serves 4***

1 Heat the oven to 200C/400F/gas 6. Drop the cauliflower florets in a saucepan containing 2.5cm/1in boiling salted water and cook for about 7 minutes, or until they are just tender; drain well.

2 Meanwhile, put the flour into a small bowl and blend to a smooth paste with a little of the soured cream, the mustard, half the grated cheese and salt and freshly ground black pepper to taste. Mix in the remaining soured cream.

3 Mix in the cauliflower and mushrooms, turning gently so that all the vegetables are coated. Pour into a shallow ovenproof dish.

4 Using a fork, mix the butter with the oats, then add the rest of the cheese and the walnuts to make a lumpy, crumbly mixture. Sprinkle evenly over the cauliflower and mushrooms and bake for 40 minutes, until the topping is golden brown.

BLUE CHEESE CHAMPIONS

SER GIOVETO

With its incisive, dominant taste, blue cheese is not as easily matched with wine as many other cheeses – but some do partner it perfectly

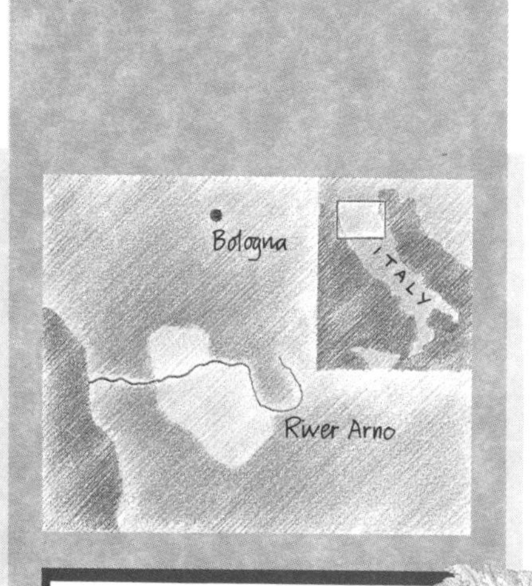

BEST BUYS

Ser Gioveto 1986 Rocca delle Macie
Taste guide: full-bodied

Montecastelli Chianti Classico Riserva 1981
Taste guide: medium-bodied

Bourgueil 1987
Taste guide: light-bodied

Madeira Malmsey Veiga Franca
Taste guide: very sweet

Bernkasteler Kurfürstlay Riesling Spätlese 1983
Taste guide: sweet

Vouvray 1987 Domaine Gauchier
Taste guide: medium-dry

IT HAS BEEN said and written so often that wine, red wine in particular, goes well with cheese that hardly anybody questions it any more. It is assumed to be a fact. So much so that the concept of wine and cheese parties has grown up; smart, one wine per course dinners are planned with an important red wine pitched against the cheese; cheese is the most common dessert for those (particularly in restaurants) wishing to finish the bottle of red they were drinking. It really is time that our preconceptions were put to the test. For cheese can be quite strongly flavoured in its own right and there certainly isn't just one entity known as 'cheese'.

Blue cheese, in particular, has a strongly incisive, dominant taste which, if you stop and think about it, is not the sort of flavour you would imagine to be well matched with fruit – and wine is fruity. Even with blue cheese there is not just one type. Just three of the most common – Dolcelatte, Danish Blue and Stilton – are very different in texture, strength, type and amount of mould, and flavour. Trying to find a wine which goes well with all of these (and therefore has a good chance of going with many other blue cheeses) is much harder than you might imagine. Some taste bitter, some metallic, some both. And some are so strongly flavoured that you can't taste the cheese at all.

Power struggle

In the end what it comes down to is that a good wine for blue cheese must have that all-important fruit, but slightly sweet-tasting fruit (implying a wine from a warmish climate); but it must also have strong acidity. For cheese is rich and fatty and, like all fatty foods, needs acidity in its partnering wine to cut through it. The third criterion is that the flavour of the wine is strong enough not to be overpowered by the powerful taste of the cheese.

Tuscany twosome

Italian red wines nearly all have good acidity – far more so than the wines of any other country; the climate is generally warm, and it also produces many richly flavoured wines. A perfect example is **Ser Gioveto 1986 Rocca delle Macie.** It is from Tuscany, from the Chianti district. Chianti is a blend of grapes;

this wine instead is made just from one, Sangiovese, the best and highest quality grape of the area. Ser Gioveto is wonderfully deep in colour, with a nose of bramble fruit and cinnamon and a concentrated, harmonious, richly fruited taste like wild blackberries and allspice. It tastes leaner and drier when partnered with blue cheeses but still delicious – and the cheese flavours come through clear and strong.

Along the same lines, also from Tuscany is **Montecastelli Chianti Classico Riserva 1981** from Castell'in Villa, one of Chianti's most respected estates. Older, softer, more mature, less assertively fruity, more gently truffley, it still has all the necessary characteristics and would be a good choice for something to go both with the main course and cheeses.

A surprisingly good blue cheese match is **Bourgueil 1987** from Pierre Breton. It's surprising because, coming from the Loire valley of northern France, it doesn't have the weight and richness that should be necessary. But it does have the acidity - plenty of it - and the lively fruit (in this case grassy, raspberry-like fruit), and it *does* partner all blue cheeses perfectly, the flavours of neither wine nor cheese being squashed or contorted.

At the end of the meal

Port may well go very well with Stilton, but only if it is an expensive, rather posh Port. With bottles at more affordable prices, it suffers the same fate as many wines: it tastes metallic, even though its sweetness helps a bit. If a fortified wine fits the bill, why not, therefore, try Madeira instead. A wine like **Madeira Malmsey Veiga Franca** has all the sweetness of port, but that essential acidity too, which cuts remarkably cleanly through the cheeses, whichever varieties you might choose, while adding definition to their tastes, and is excellent value.

The suitability of the match with Port and Madeira, both sweet wines, should give the clue to a whole group of blue cheese partners, far too often ignored: white wines. German wine lovers will have no problem in drinking their favourite wines with blue cheeses; they are full of sweet fruit, have good acidity, and lively flavour. **Bernkasteler Kurfürstlay Riesling Spätlese 1983** is an ideal example.

An alternative comes from the Loire valley. **Vouvray 1987 Domaine Gauchier** has the ideal sweet-sour, floral-fruity character that is the perfect foil for blue cheeses. The sweetness isn't too obvious and the cheese tones down the acidity. The partnership is starry and the wine good value, too.

The best blue cheeses are delicious just with the wine - although a few fresh, sweet grapes or a stalk of celery make good accompaniments. If you prefer to eat cheese with biscuits, then choose a plain, crisp variety such as water biscuits, which are very easy to make (see below).

PERFECT PARTNERS

Water biscuits

Water biscuits have been a firm favourite to eat with cheese for many years

- ***Preparation: 30 minutes***
- ***Cooking: 10 minutes***

50g/2oz butter or margarine, plus extra for greasing
225g/8oz flour
pinch of salt

- ***Makes 16***

1 Heat the oven to 200C/400F/gas 6 and grease a large baking tray.

2 Sift the flour and salt into a bowl. Use a round-bladed knife or palette knife to cut the butter or margarine into the flour until all pieces are pea-sized and coated with flour. Rub the fat into the flour until the mixture resembles breadcrumbs.

3 Make a well in the centre of the mixture and pour in 6tbls cold water. Using a fork, draw the dry ingredients into the water until the mixture sticks together. Knead gently into a smooth ball.

4 Roll out 5mm/¼in thick. Use a 6.5cm/2½in cutter to cut out the biscuits, re-rolling as necessary. Then, using a rolling pin, gently roll each biscuit out to thinner circles about 9cm/3½in diameter. Prick with a fork.

5 Bake just above the centre of the oven for 10-12 minutes or until golden and puffy. Leave the biscuits to set on the tray for 5 minutes, then transfer to a wire rack to cool.

WINES WITH BRIE

COTES DU JURA

Though red wines traditionally accompany Brie, some whites complement it most deliciously

BRIE, LIKE MOST other cheeses, is usually eaten with red wine. But the only real justification for doing so, apart from for those who are not that keen on white wine, is when the Brie is part of a cheese board that includes other, more red-partnering, cheeses. For although Brie will go quite well with a number of reds it usually goes much better with white wine. After all, Brie is very rich – creamy and buttery – and there is nothing better than a crisp white to cut through such richness. It is also somewhat salty and can be a little chalky – either from the skin or, occasionally, from not-quite-ripe parts. Ripe, rounded white wine-type fruit is a far better bet than red for balancing these elements.

Enhancing the flavour

Brie, the pasteurised type at least, can be quite mild in flavour so you might think it would be overwhelmed by a strongly-flavoured wine. But its taste is quite persistent, so if you take a bite of it after a sip of wine it will often start to come through after a second or so. More important is the effect the cheese has on the wine. Some wine goes bitter or leathery, some appears to lose its fruit, but some is quite splendid. The essence of a good match is the contrast between the two tastes; they complement each other but stay distinct.

To see this match at its best try **Côtes du Jura Chardonnay 1987.** Wines from the Jura, in the far east of France, are not seen all that often. Occasionally they are rather sherry-like but this one, from the much-loved Chardonnay grape, is both rounded and crisp. Its bouquet is quite nutty, of both walnuts and hazelnuts, with greengage, buttery, biscuity aromas. It is broad and rich, fully fruited but with a good bite of acidity, mouth-filling alcohol and an enlivening tangy finish. The flavour of the Brie comes through cleanly and well but when the wine is used to clean the mouth after the cheese the two tastes seem to leap into perspective. It is rather like having an enlivening sauce on a gently-flavoured ingredient. The wine is powerful enough too to follow on after a red if the cheese is eaten at the end of a meal.

Californian choice

Another Chardonnay that is a good Brie partner and is also weighty

BEST BUYS

Côtes du Jura Chardonnay 1987
Taste guide: **dry**

Belvedere 1987 Sonoma County Chardonnay
Taste guide: **dry**

Verdicchio dei Castelli di Jesi Classico 1987 Garofoli
Taste guide: **dry**

Château Bellevue La Forêt Côtes du Frontonnais
Taste guide: **medium-bodied**

Malmsey Madeira Donaldson
Taste guide: **very sweet**

enough to follow red wines or strongly flavoured foods is **Belvedere 1987 Sonoma County Chardonnay.** It is from California, but from the cooler coastal part well north of San Francisco. Nevertheless it is still warm enough for the grapes to ripen well and for the wine to develop a big, warming, toasty, buttery richness. So rich is it that it appears a touch sweet but it is not too full-bodied, so very drinkable. Were it any more powerful there would be a risk that very mild Brie might be overwhelmed, but instead the balance is just right, with both cheese and wine filling the mouth with their flavours and providing a good contrast to each other's taste.

Is white right?

If the cheese forms the main part of the meal in its own right, perhaps as a light lunch or supper or at a 'cheese and wine' party there is no need to have such a fully-flavoured wine. Indeed, you may prefer something much lighter in style. A good choice would then be **Verdicchio dei Castelli di Jesi Classico 1987 Garofoli.** Verdicchio is usually thought of as one of Italy's best wines for fish and sea food, although it will partner white meats too. Its affinity with Brie is a pleasant surprise. The wine takes its name from the castles in and around the small town of Jesi, near Ancona on Italy's Adriatic coast. The wine smells of hedgerows and the sea – the apparent (not real!) saltiness perhaps giving a hint of how well it will match the slightly salty Brie. Its peachy, appley fruit has a roundness that hides its cutting acidity – which is so necessary to cut through the fat of Brie cheese.

Of course, you may decide that the wine just has to be red, or you may be serving the cheese as part of a buffet, say, where people will be free to choose either red or white. You will need a red that is not too big and has plenty of fruit. It shouldn't be too tannic, especially if it has good acidity too, or else the cheese can taste of ammonia! Go for **Château Bellevue La Forêt** from south-west France's Côtes du Frontonnais. It has a strongly damson-like bouquet, and a lively, firm fruitiness with a crisp, dry finish. For a more vibrant wine choose the 1988, for a more rounded, mellow style, get the 1987.

Dessert wine

If the Brie comes at the end of a more important meal and you fancy a big, strong wine with it, forget Port and try **Malmsey Madeira** from Donaldson. It has a fabulous red-brown caramel colour and smells richly toffee-like and very inviting. It is sweet but with a wonderful intensity of smoky, tangy, toffeed flavours and – most importantly – a good swipe of acidity (which Port lacks). It is this acidity which makes it such a good complement to Brie. Although the sweetness of the wine coats the mouth and it takes a moment or so for the cheese's taste to come through, the two tastes balance perfectly and it turns into a gloriously hedonistic partnership.

If you prefer to serve a selection of cheeses, rather than just one, then choose two or three cheeses with contrasting flavours, colours and textures. A firm, well-flavoured cheese such Red Leicester or a mature Cheddar would be a good bet, along with a slightly softer cheese such as Chaumes, or Bel Paese.

PERFECT PARTNERS

BRIE

One of the favourite cheeses for any cheeseboard is Brie, sometimes called 'the king of cheeses', and certainly one which pleases even those who avoid the strong flavour of ripe Camembert. Good grocers and delicatessens sell it in thin yellow crusted cartwheels dusted with a powdery white mould. Ideally, it should be eaten when it is soft and slightly runny, but if it is allowed to go past its best it can have a strong smell which affects the flavour adversely. Under-ripe, it is apt to be chalky, so buy with care. When it is not sold by weight, it comes in wedge-shaped boxes. Delicious with crusty bread, it is equally good with a selection of biscuits for the cheeseboard – Bath Olivers, oat-cakes, water biscuits and crispbread. And a small bunch of grapes or a crisp apple make a welcome accompaniment! Or try it with Spanish quince cheese *(dulce de membrillo)*.

PASS THE PORT

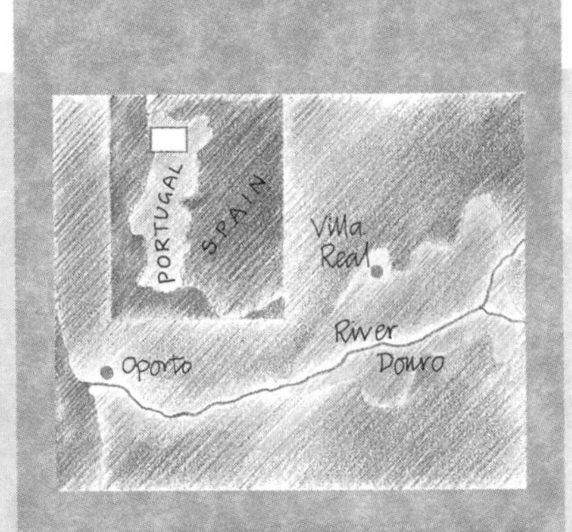

OFFLEY BOA VISTA 1982

A warming glass of port for that after-dinner drink

THE IMAGE OF Port is a group of middle-aged gentlemen, comfortably well-fed, rather solemnly and respectfully passing a decanter of strong, dark, red wine – from right to left of course, pouring a small glassful and reverently tasting and drinking it. It still happens from time to time, in certain circles, but Port can be enjoyed without all the ceremony and ritual, even without decanting. The only Port that must be decanted is Vintage Port. Unlike most other wines where vintage means simply the year of harvest, only the best wines from the best years are declared Vintage by the Port producers. Vintage Port, deep in colour, strong and sweet, is bottled at between two and three years old – that is young in Port terms and takes about twenty to become ready to drink. In that time a thick deposit or 'crust' develops in the bottle so careful decanting is essential to avoid bits of black sludge in your glass. Vintage Port is also horribly expensive and therefore only for the most special of occasions. Thank goodness there are several other types of Port which are more affordable.

'Vintage'

Most Ports that mention 'vintage' on a label are either Vintage Character or Late Bottled Vintage. Vintage Character is a *very* rough approximation to real Vintage. The main similarity is that it is bottled young, while it is still a deep ruby colour. It is much less intense in flavour, though, and it doesn't need to be kept, so won't develop a deposit and can therefore be poured straight from the bottle without decanting. A good example is **Smith Woodhouse Vintage Character** with its sweet, redcurrant, medicinal bouquet and its alcoholic, tangy, concentrated fruit sweetness.

Late Bottled Vintage is similar, but of somewhat higher quality and bottled at four to six (usually nearer six) years old. It *does* come from wine of one year's harvest but, unlike true Vintage, the harvest doesn't have to be an exceptionally good one and (except for the rare examples bottled at four years old) it won't need decanting. The year of harvest is shown prominently on the bottle and, in smaller print, the year of bottling too. It should be more concentrated than Vintage Charac-

BEST BUYS

Offley Boa Vista L.B.V. 1982
***Taste guide:* very sweet/ full-bodied**

Smith Woodhouse Vintage Character
***Taste guide:* very sweet/ full-bodied**

Reserve 10 Year Old Tawny Port (Royal Oporto Wine Company)
***Taste guide:* very sweet/ full-bodied**

Warre's Warrior
***Taste guide:* very sweet/ full-bodied**

ter and smell less obviously jammy sweet, more bracken-like and spicy. It should also taste a little rounder and more mellow. Unfortunately some Late Bottled Vintage, or L.B.V. as it is often labelled, is hardly better than Vintage Character. Of the good ones, **Offley Boa Vista L.B.V. 1982** (bottled 1988) shines. It tastes just as it should and is particularly harmonious, without any excess spiritiness that is sometimes present.

Ruby and Tawny

The ruby colour of young bottled Port has spawned the name Ruby for any Port of this style. The longer Port spends in the cask the lighter and less red it becomes, turning a tawny hue. Hence the name Tawny Port. But beware! Port sold just as 'Tawny' is not a true, wood-aged Port, but a young one: young Ruby and White Ports are mixed to get the right colour! Most *real,* wood-aged Tawny is sold as '10 Year Old Tawny'. This means the wine has been aged an average of approximately ten years in casks before bottling. 10 Year Old Tawny is a very different sort of wine from the young-bottled Ports that are more common. Apart from the colour, a light red-brown, the bouquet is much softer and more nutty than fruity, while the taste is not at all fiery but gently nutty and raisiny. A lot of true Port *aficionados* prefer this sort of aged Tawny even to the giddy heights of real Vintage, and with its gentle mellowness it is easy to see why. Try **Reserve 10 Year Old Tawny Port** from the Royal Oporto Wine Company and you will have a fine example. There is no need to decant it.

Much as ordinary Tawny Port bears little resemblance to real, aged Tawny, so ordinary Ruby Port is very ordinary indeed compared with the better types, whether they call themselves 'Vintage Something' or not. After all, even Vintage Port itself is a special example of a Ruby. Life is not made any simpler by words like 'Fine' or 'Finest' or other terms denoting special quality on bottles of ordinary Ruby: they have no significance at all and are no more than meaningless puff. This means that the term Ruby has become debased, so much so that some Port houses which make particularly good wine of this type have decided to ignore the name altogether. **Warre's Warrior** is one such. Strong and intensely rich, packed full of flavour, with plum and redcurrant fruit, and plenty of the rather medicinal twang that signifies quality, it has more of the character of Vintage about it than most of the others of this type.

Keeping qualities

As Port is very strong in alcohol (20%) a small glass will go a long way. And it keeps quite well once opened, so although a bottle may seem expensive it can last for a number of meals.

If you like, serve a selection of petit fours with the Port – fondant fruits or chocolate-dipped citrus peel can be bought in good delis, or you could make some rolled cigarette russes (see below) to accompany them.

PERFECT PARTNERS

Cigarettes russes

The secret of success with these elegant, black-tipped 'Russian cigarettes' is to cook them one at a time to ensure that they are soft for rolling

- ***Preparation: 40 minutes***
- ***Cooking: 6 minutes***

2 large egg whites
100g/4oz icing sugar, sifted
40g/1½oz flour
65g/2½oz butter, melted
a few drops of vanilla essence
25g/1oz dark chocolate
butter and flour for greasing

- ***Makes 15-20***

1 Heat the oven to 200C/400F/gas 6 and grease and flour 2 baking trays.

2 In a bowl whisk the egg whites until stiff. Sift the icing sugar onto them and gently beat it in with a wooden spoon.

3 Sift the flour onto the mixture. Beat it in carefully making sure the mixture is well combined. Beat in the butter.

4 Drop 1tbls of the mixture on one of the trays. Smooth out thinly and evenly with a knife. Bake for 5-6 minutes until the biscuit is set and golden brown. Repeat on the other tray halfway through the cooking of the first biscuit.

5 Run a palette knife or fish slice under one cooked biscuit and turn it upside down. Quickly roll it up around a pencil. After a few seconds when each biscuit has hardened, remove the pencil and cool the biscuit on a wire tray.

6 Cook the remaining mixture in pairs in the same way.

7 Melt the chocolate in a bowl over a saucepan of hot water. Dip one end of each biscuit into it. Use the end of a wooden spoon to clear the chocolate from the centre of the biscuits. Put the biscuits back on the wire tray until the chocolate has cooled. Store in airtight container in a cool place.

SWEETS FOR YOUR SWEET

CHATEAU DE BERBEC

A love of sweet things and a fondness for wine mean you're in for a treat if you have a dessert wine with your pudding

SELECTING THE RIGHT sweet wine to partner a dessert is surprisingly difficult. The problem is judging the level of sweetness. If the food is sweeter than the wine, the wine will taste sharp; if the wine is sweeter than the food, the food will taste sour. Even if the sweetness is more or less balanced, a rich wine served with a delicate dessert can be cloying or the flavours of a light wine can be squashed by a fully flavoured pudding. But get the right match and sweet wine can make an ordinary dessert seem special or add finesse to a special dessert.

The most well-known dessert wine is Sauternes. Sadly, its image became tarnished in the sixties and seventies when the misspelled 'Sauterne' was used to describe any sweet wine, however cheap and nasty. Real Sauternes is something else. It comes from the Bordeaux region of France and is made from Sémillon and Sauvignon grape varieties. To make it sweet, the grapes are left on the vines long after the normal vintage time. With the hoped-for sunny days and misty mornings of autumn, a special mould grows on the grapes. The mould extracts the moisture, thus concentrating the sugars and severely reducing the yield. The result is grapes so rich in sugar that the yeasts can't ferment the wine to dryness, and the Sauternes' special taste.

The riskiness of the process (if the weather's bad the mould won't arrive and the crop is lost) and the tiny quantities resulting mean that good Sauternes isn't cheap. As its taste is full and its price quite high you won't want a lot, so half-bottles are ideal. To taste the real thing without spending a fortune there's the simply named **Vintage Selection Sauternes 1984.** But scan the back label and you will find it is really **Cypres de Climens**, the wine made by the starry Château Climens from the grapes which are almost but not quite good enough for the expensive Château wine. Its honeyed, barley-sugar flavour, with lots of sweetness — but also freshness — is remarkable. Sauternes is often recommended to be drunk with no more than 'a fine fresh peach', but any not over-sweet fruit-based dessert should do the trick.

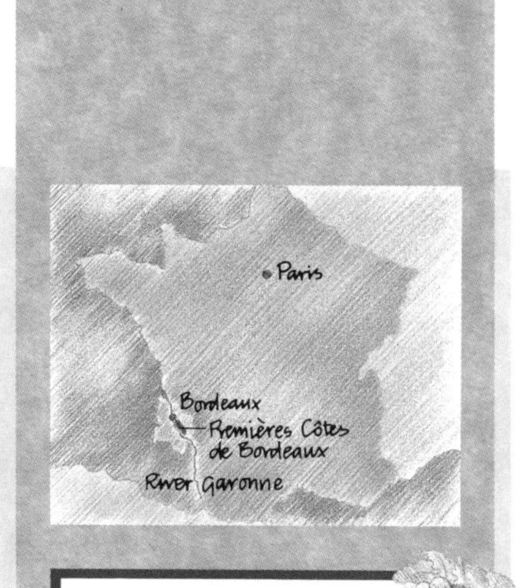

BEST BUYS

Château de Berbec 1985 Premières Côtes de Bordeaux (two glasses)
Taste guide: **very sweet**

Cypres de Climens (two glasses) **(half bottle)**
Taste guide: **very sweet**

Moscato Frizzante (one glass)
Taste guide: **sweet**

Moscatel de Valencia (one glass)
Taste guide: **very sweet**

Sauternes substitute

For an excellent Sauternes substitute there's **Château de Berbec 1985 Premières Côtes de Bordeaux.** Premières Côtes de Bordeaux is not far from the Sauternes district, just across the river Garonne, so the wines are often similar in style, though rarely as refined. Château de Berbec is bright gold and has a beautiful honeyed lanolin nose (the Sémillon grape often smells like lanolin). Although sweet, it's packed full of strong, rich, honeyed fruit flavour — almost good enough to be a liquid dessert in its own right.

To solve the problem of conflicting sweetnesses of food and wine there's Asti Spumante from Italy. Asti combines extreme lightness of weight, plenty of sweetness, good acidity and low alcohol, which is a remarkably good combination for dessert partnering. Asti is the perfect foil to a surprising number of puddings, be they rich and heavy (even Christmas pudding) or light and airy (like soufflés). Practically any Asti Spumante will suit. But for something a little different why not try a **Moscato Frizzante?** Moscato (or Muscat) is the grape from which Asti is made. It's a uniquely flavoured grape, as good to eat as to make into wine, and the only grape whose wine tastes truly grapey. Instead of making a fully sparkling wine, as they do around the town of Asti, Moscato Frizzante is only slightly sparkling but keeps all the lovely Muscat flavours — which are easier to taste without all those bubbles! Light in body and alcohol, cool and refreshing, it will enliven many a rich dessert.

Alcohol and oranges

Muscat is grown in many countries of the world and each produces wine of a totally different character; sometimes dry, sometimes sweet, occasionally as light as Asti, or maybe rich, fat and heavy. In Spain the grape is called Moscatel. Richer than its Italian cousin but not too weighty, it often takes on an aroma of oranges. **Moscatel de Valencia** is a prime example. It certainly has a grapey-orangey bouquet. Its taste is quite perfumed with a good balance between sweetness and alcohol — both quite high but neither dominating — and that orange character again, all with a smooth silky feel in the mouth. It's definitely a winner with any dessert using oranges.

CHILLING THOUGHTS

Sweet wine should be served well chilled, so keep it in the fridge for a good while – at least an hour – before serving. Bring the wine to the table, straight from the fridge, a few minutes before you need it. Pour it at once and let it warm a little in the glass. As the aromas start to emerge, try your first sip – it should be delicious!

PERFECT PARTNERS

Summer pudding

Remember to prepare this delicious pudding the day before it is needed.

- ***Preparation: 30 minutes, plus chilling overnight***
- ***Cooking: 10 minutes***

100-175g/4-6oz sugar
100g/4oz redcurrants
100g/4oz blackcurrants
450g/1lb cherries, stoned
225g/8oz raspberries
8-10 large slices of white bread, about 5mm/¼in thick, crusts trimmed off
whipped cream, to serve

- ***Serves 4-6***

1 Put the sugar in a heavy saucepan with 150ml/¼pt water and place over low heat until the sugar has completely dissolved, stirring constantly. Add the fruit, stir very gently and simmer for 5 minutes. Leave to cool, then drain the fruit, reserving the syrup.

2 Sprinkle an 850ml/1½pt pudding basin with water. Cut a round piece of bread to fit the bottom of the bowl and lay it in place.

3 Line the sides of the bowl with some of the pieces of bread cut in half lengthways, trimming as necessary to fit the bowl exactly. Reserve the remaining slices of bread.

4 Fill the bread-lined bowl with the fruit and cover with all the reserved pieces of bread.

5 Select a flat plate or saucer which fits exactly inside the rim of the bowl and weight it down well. Chill the pudding and reserved syrup overnight.

6 Turn the pudding out onto a flat serving dish and pour over the reserved syrup. Serve with whipped cream.

WINES FOR HALLOWEEN

CHATEAU DE JAU

Evocative wines for keeping ghosties, ghoulies and witches on broomsticks at bay

IT IS THE night of ghoulies and ghosties, witches and warlocks, and you may feel the need for a good shot of something deeply coloured, full-bodied and powerful to keep the spirits at bay. Or you may be celebrating their evening with hollowed-out pumpkins, apple-bobbing and the like and need some wine to help the evening along. Either way you will need some wines for Halloween.

Red or white?

It should be red wine for best protection against the black witches and solidarity with the white. But there is no point in drinking red wine if you enjoy white more – you will just have to have a white even the witches will respect, like **Chardonnay de Chardonnay 1988.** It is from the Mâcon district of Burgundy and from the Chardonnay grape variety. But the Chardonnay is as undefiled as it could be as the vines grow around the village, also called Chardonnay, that gave its name to the grape. Not only will the evil spirits fizzle away under the assault of such virginal purity, the wine tastes good for mortals too. It is light, crisp, salty, creamy and biscuity; fine on its own but with enough weight to cope with Halloween party fare or a fish or white meat course at dinner.

A good bet for a red is **Château de Jau 1985** if for no other reason than it evokes the warmth of its origins and is soft and comforting enough to reassure you against whatever shadows may be lurking outside. The warmth is that of the south of France, the Côtes du Roussillon in the south west. The deep but bricky red colour, the ripe, dry-earthy, stony, minerally, herby, peppery fruity taste all contribute to a rounded, full mid-weight good wine which is equally at home with sausages on sticks, crisps and fried mushrooms in breadcrumbs, as with roasts, sautées and mixed cold meats.

Big flavours from Chile

Are there witches in Chile too? Halloween would be summer there, so there would be few hours of darkness to harbour them. Either way, a **Chilean Cabernet Sauvignon** is just what you need for a cold,

France
Perpignan
River Têt
France
Spain

BEST BUYS

Château de Jau 1985
Taste guide: medium bodied

Chardonnay de Chardonnay 1988
Taste guide: **light and dry**

Chilean Cabernet Sauvignon
Taste guide: **full bodied**

Vino Nobile di Montepulciano 1985
Taste guide: **full bodied**

Mavrodaphne of Patros
Taste guide: **very sweet**

dreary end of October whether you believe in the spooks or not. It is a deep ruby red and has a big, ripe aroma full of blackcurrants – just like the Cabernet grape should. Dry, ripe and rounded, not too tannic, with plenty of blackcurrant flavours too and a little mineralliness, it will feel warming and protective. It will go with foods as robust as casseroles, pies, duck and roast poultry dishes, chops and will partner mild cheeses and savoury nibbles too.

If you are looking for powerful stuff to keep away the ghosties, but you want plenty of taste as well, then go for the robust **Vino Nobile di Montepulciano 1985 Poliziano**. Poliziano is one of the best producers in the small area called Montepulciano in southern Tuscany in central west Italy. (The area has nothing to do with the Montepulciano grape, which is grown elsewhere.) 1985 was one of the best years of the 1980s and has produced big, gutsy, long-lasting wines. This means that this Vino Nobile still has lots of tannin which makes it too harsh to drink on its own. But if you pour it into a jug (or decanter if you have one) about twenty minutes before you drink it, this will help it to soften. If you drink it with succulent steaks and other grilled red meats or hearty red meat dishes you will find the tannin hardly noticeable. What you will taste will be the delcious, mouth-watering plums, cherries and cinnamon-tinged fruit. Not for the apple-bobbers, this one, but it is great for those enjoying the evening with a good meal. Even the witches might be envious of your good fortune to be drinking such a bottle.

Halloween special

Mavrodaphne of Patros is a Halloween special. It is strong (15%), red, smells a bit porty, and is sweet, with a rich, blackberry, slightly almondy, almost honeyed sweetness that encourages you to drink it - and come back for more. You couldn't serve it throughout the evening - it is far too alcoholic for that - but a glass with dessert, cheese, coffee, petits fours and nuts will be a splendid and apt end to a Halloween dinner. Similarly, if your Halloween party has any sweet dishes, serve Mavrodaphne with them, or leave a bottle by the cheese tray. When the witching hour itself arrives, give everyone a glassful of this wonderfully rich spirit-beater.

And while the adults are sipping wine and munching on nibbles and nuts, serve the children a truly terrific Halloween special - trick or treat toffee apples (see below) are simple to make and quite delicious to eat - some of the adults might even be tempted to forego the wine!

PERFECT PARTNERS

Trick or treat toffee apples

- ***Preparation: 10 minutes***
- ***Cooking: 15 minutes***

8 small dessert apples, stalks removed
225g/8oz soft dark brown sugar
25g/1oz unsalted butter
1tbls golden syrup
1tsp malt vinegar

- ***Makes 8*** (ff) (££)

1 Push a wooden skewer through the centre of each apple, and lightly grease a baking tray.

2 Put the sugar, butter, syrup, vinegar and 5tbls water into a saucepan and stir over low heat until the sugar has dissolved. Bring to the boil and cook until the temperature of the toffee reaches 143C/290F; at this temperature a small amount of the mixture dropped into cold water separates into threads and becomes quite hard.

3 Remove the pan from the heat and quickly dip each apple into the toffee. (Do this carefully as toffee is very hot.) Turn to coat it completely. Then place in a bowl of cold water so that the toffee sets hard.

4 Lay the apples on their sides on the baking tray until the toffee cools. Wrap in cling film until ready to serve.

WINTER WARMERS

MALMSEY

If it's cold, grey and miserable outside, a glass of Port, Madeira or sherry might be just what you need to warm up

FOR A START it has to be said that alcohol won't really warm you up. In fact it's the reverse: it might make you feel warm but will actually cool you down, so make sure you're settled by the fire or in a warm room first. A little glass of something big seems the most tempting prospect; strong and a bit sweet —and the most obvious answer is Port.

Passing the port

Port comes from northern Portugal's Douro valley from vines grown on precipitous rocky terraces above the winding river. It comes in a wide variety of styles and qualities and a great range of prices, but the major distinction is how long it has been kept in wood: large old oak barrels. Port that is bottled after just a few years is still a bright ruby red colour (hence the name Ruby Port), vibrant, fruity and sometimes fiery. The best come from one year's harvest only, are bottled after just two or three years and called Vintage Port. They need to mature in bottle for many more years before being softened enough to drink and they're horribly expensive.

Wines in the style of Vintage Port but less intense and usually drinkable as soon as bottled are called Late Bottled Vintage or Vintage Character. **Smith Woodhouse Vintage Character** typifies the hearty fruitiness of such styles and is good value. A bright ruby colour, its bouquet is truly porty: wine gums and old fashioned pharmacists' shops. Strong, but not too spirity, the sweet, ripe, luscious fruit wraps itself round the mouth in the most comforting of ways and the feeling of warmth trickles slowly down the throat and stays there.

Taylor's First Estate is also a vigorous, young-bottled and fruity Port, but more concentrated, more mellow - and more expensive. A sniff reveals some wood along with the pleasantly medicinal aromas, not over obvious but very powerful. Less overtly sweet but big, ripe and harmonious, a glassful, possibly with a few nuts or a nibble of hard cheese, will make an hour or so drift by in warm contentment.

BEST BUYS

Cossarts' Over 5 Years Reserve Finest Malmsey
Taste guide: ***very sweet***
Serve with: ***mince pies, Madeira cake, raisins***

Smith Woodhouse Vintage Character
Taste guide: ***sweet***
Serve with: ***unsalted nuts, strong hard cheese***

Taylor's First Estate
Taste guide: ***sweet***
Serve with: ***hard cheese***

Warre's Nimrod
Taste guide: ***sweet***
Serve with: ***unsalted nuts, milder hard cheeses***

Harvey's '1796' Palo Cortado Sherry
Taste guide: ***dry***
Serve with: ***crisps, nuts and other salty nibbles***

The other major type of Port is kept to mature much longer in wood, until its colour fades to a red-brown, tawny hue. Called Tawny Port, it is softer, gentler and nutty rather than fruity and can be anything from about eight to over 40 years old. True Tawny can't be cheap; it takes so many years to produce it. A little, though, goes a long way. **Warre's Nimrod** is hard to beat for a sprightly rather than venerable old Tawny. Hearteningly rounded, nicely nutty, with its alcohol well integrated into a calm whole, it will gently soothe on the iciest of days.

A word of warning: anything sold just as Tawny Port or Ruby Port is a skilled blend created to imitate the characteristics of these types - cheaper because less aged, but consequently less good.

Moving to Madeira

But a winter warmer needn't be Port. Best of all, and most cheering, is Madeira. From the island of the same name, it comes in four major styles: Sercial, Verdelho, Bual and Malmsey, in increasing order of sweetness. Its flavour is unique because the wine is made by a special process never seen elsewhere. The wine is gradually heated to around 40-50C/104-122F, kept there for a few months, then allowed to cool gradually. This gives it a delicious, caramelized tang. It replicates how the wine was originally created: it was carried as ballast on trading ships heading for the Far East across the equator, where they were often becalmed. The heating, called *estufagem*, also stabilizes it so Madeira, once opened, will last far longer than any other wine. **Cossarts' Over 5 Years Reserve Finest Malmsey** is Malmsey at its most enticing. Bright, dark amber-brown, it smells a little of salt and cheese as well as caramel. Its taste is the liquid equivalent of a boozy rich dark fruit cake: not only warming but wonderfully self-indulgent.

Staying with sherry

If you want to enjoy the invigorating effects of such fortified wines but don't fancy anything sweet there's **Harvey's '1796' Palo Cortado Sherry.** A Palo Cortado is a rare type of sherry, somewhere between an Amontillado and an Oloroso in style. '1796', though, has been long aged in wood, gaining body and depth so its roundness comes from its concentration, not sweetness. Russet-coloured, a bouquet and taste of marinated walnuts and a big, firm, tangy mouthful, it will chase away the damp and the chill — and stimulate the appetite.

Sweet or dry, sherry is the ideal accompaniment to many foods: a dry sherry is perfect with olives, nuts, spicy sausages, mushrooms or kidneys, and a sweet one can accompany nuts, cheese or pudding.

PERFECT PARTNERS

Madeira cake

This cake acquired its name because, in the 19th century, it was often served mid-morning with a glass of Madeira

- ***Preparation: 20 minutes***
- ***Cooking: 1 hour***

175g/6oz butter, softened, plus extra for greasing
175g/6oz caster sugar
3 large eggs, beaten
100g/4oz plain flour
100g/4oz self-raising flour
1-2tbls milk
2 slices candied citron peel (optional)

- ***Serves 6-8***

1 Heat the oven to 170C/325F/gas 3. Grease the sides and base of a deep 18cm/7in round cake tin. Line with greased greaseproof paper.

2 Cream the butter and sugar in a large bowl until light and fluffy, then beat in the eggs, a little at a time. Sift together both flours and lightly fold into the mixture with a metal spoon, adding enough milk to give it a soft dropping consistency.

3 Spoon the mixture into the tin, level the top, then make a slight dip in the centre. Bake for about 1 hour, or until well risen and golden brown. You can tell if it is cooked by inserting a fine skewer in the centre; it should come out clean. Place the citron peel on top halfway through the baking time, if using. Remove from the tin, peel off the lining paper and invert onto a wire rack to cool completely.

INDEX

Acknowledgements

The publishers extend their thanks to the following agencies, companies and individuals who have kindly provided illustrative material for this book. The alphabetical name of the supplier is followed by the page and position of the picture/s. Abbreviations: b = bottom; c = centre; l = left; r = right; t = top.

All pictures from the MC Picture Library: Sue Atkinson: 48br; Bryce Attwell: 126br; Tom Belshaw: 84b; Martin Brigdale: 100r, 104br; Paul Bussell: 58br, 82br, 90b; Chris Crofton: 122br; Ray Duns: 74br, 78br, 132br; Laurie Evans: 62br, 70br, 88br, 116bl; Edmund Goldspink: 20b; John Hollingshead: 128br; James Jackson: 16br, 68br, 102br, 120br; Chris Knaggs: 12br, 32b, 38br, 98br, 118r; Jess Koppel: 18br, 24br, 124br; Michael Michaels: 142br; Peter Myers: 34br, 36br, 50br, 52br, 76br, 94br; Roger Philips: 66br, 134br; Ian Reid: 96bl; Grant Symon: 72br; Paul Webster: 54br; Andrew Whittuck: 138br; Paul Williams: 22b, 28br.